T0205391

Studies in Big Data

Volume 118

Series Editor

Janusz Kacprzyk, Polish Academy of Sciences, Warsaw, Poland

The series "Studies in Big Data" (SBD) publishes new developments and advances in the various areas of Big Data- quickly and with a high quality. The intent is to cover the theory, research, development, and applications of Big Data, as embedded in the fields of engineering, computer science, physics, economics and life sciences. The books of the series refer to the analysis and understanding of large, complex, and/or distributed data sets generated from recent digital sources coming from sensors or other physical instruments as well as simulations, crowd sourcing, social networks or other internet transactions, such as emails or video click streams and other. The series contains monographs, lecture notes and edited volumes in Big Data spanning the areas of computational intelligence including neural networks, evolutionary computation, soft computing, fuzzy systems, as well as artificial intelligence, data mining, modern statistics and Operations research, as well as self-organizing systems. Of particular value to both the contributors and the readership are the short publication timeframe and the world-wide distribution, which enable both wide and rapid dissemination of research output.

The books of this series are reviewed in a single blind peer review process.

Indexed by SCOPUS, EI Compendex, SCIMAGO and zbMATH.

All books published in the series are submitted for consideration in Web of Science.

Aboul Ella Hassanien · Ashraf Darwish
Editors

The Power of Data: Driving Climate Change with Data Science and Artificial Intelligence Innovations

 Springer

Editors
Aboul Ella Hassanien
Faculty of Computers and Artificial
Intelligence
Cairo University
Cairo, Egypt

Ashraf Darwish
Faculty of Science
Helwan University
Cairo, Egypt

ISSN 2197-6503   ISSN 2197-6511 (electronic)
Studies in Big Data
ISBN 978-3-031-22458-4   ISBN 978-3-031-22456-0 (eBook)
https://doi.org/10.1007/978-3-031-22456-0

This Springer imprint is published by the registered company Springer Nature Switzerland AG
The registered company address is: Gewerbestrasse 11, 6330 Cham, Switzerland

Preface

The UN COP27 2022 conference in Egypt on climate change offers opportunities for leaders of the world to take action and make meaningful and urgent commitments and strategies to reduce greenhouse gas emissions to control global temperatures. The climate change impact is already threatening human existence across the universe. Climate change is one of the humanity's major challenges in the current century. It will continue to be a major problem in the next century not only due to its long-term effects on sustainable development, its pervasive nature, complexity, and consequences such as environmental hazards such as drought, desertification, and healthcare and agricultural problems. Climate change is used to describe the alteration in the climate over time caused by human activities and some naturally stirring events such as earthquakes and explosions. As a result, global warming is caused by the variabilities in climate variables which are occasioned mostly by human activities. Several current research studies have predicted that if immediate action is not made to address the impacts of the changes in the climate, the implications could lead to a lack of food and water in the next decades, as well as the effect on the health of people.

As a result, the current situation necessitates a quick response from all parties involved in devising, adopting, and implementing the essential steps and adaptation for reducing these potential hazards through data science and emerging and innovative technology.

Finding sustainable solutions to this issue will significantly affect the global, regional, national, and local community levels. The urgency in finding technological solutions to this global problem will require effective actions from governments, the industrial level, the private sector, and civil society.

There is an increasing recognition that data science and emerging and innovative technologies will play a major role in the national and global strategies to combat climate change challenges. Therefore, these technologies can potentially present solutions to complex development challenges brought by climate change. Data science and emerging digital and innovative technologies such as artificial intelligence, the Internet of things, digital twins, blockchain, drones, cloud computing, and sensor networks have drawn significant attention in the last years to tackle climate change challenges and reduce greenhouse gases and their impact. The impact and deploying these technologies on human life have witnessed a great interest in some areas related to climate change adaptation and mitigation, such as renewable energies, ocean dynamics, environmental quality control, biodiversity, agriculture, water sustainability, and energy sustainability. In addition, emerging digital technologies can deliver more efficient, rapid, and reliable decision-making solutions before, during, and after the occurrence of hazards as a consequence of climate change. These technologies can provide a chance to deliver sustained solutions to many societal challenges relating to climate change. Therefore, data science and emerging and innovative digital technologies can improve resilience to global warming-related, natural hazards, reduce emissions, and enhance humans' ability to take the necessary steps to realize net zero.

The contents of this book are divided into four parts.

Part "Artificial Intelligence in Climate Change Applications" is devoted to the applications of artificial intelligence for floods crises prediction based on machine learning techniques, the impact of climate change on air flight emissions for green aviation, the prediction of water quality index using machine learning techniques, the impact of artificial intelligence on waste management for climate change, and finally, a machine learning-based model for predicting temperature under the effects of climate change.

Part "Emerging Technologies in Industry and Energy Sector" concerns are emerging technologies' role in the climate change industry and energy sector. It includes the prediction of emissions of cars based on artificial intelligence techniques, a decision support framework for photovoltaic renewable energy prediction, a clean energy management system based on the Internet of things and sensor networks, digital twins for the energy management sector, using a case study based on Internet of things to design ozone prediction model for mitigating the effect of climate change, and finally suing emerging technologies for sustainable zero energy in buildings using multistage optimization in Egypt.

Part "Emerging Climate Change Technology in Agriculture Sector" is devoted to presenting the applications of emerging technologies in the agriculture industry. It concludes the crop prediction with climate factors and soil properties based on an intelligent recommendation mode, an intelligent recommendation model for strategic crops in Egypt using deep learning, a data-driven decision support system for an innovative water system for climate change problems, and finally, the role of artificial intelligence in water management systems for irrigation in agriculture.

Part "Emerging Climate Change Technologies in Healthcare Sector" presents the application of emerging technologies in the healthcare sector to tackle climate change problems. It includes the influence of climate change on the re-emergence of malaria using artificial intelligence.

The authors of this book wish to acknowledge the encouragement of the organizing committee of COP27, who motivated them to prepare this book. The success of publishing this book is first of all a result of the quality and the motivation of its authors and researchers. So, we would like to thank all participants of COP27 and researchers and authors of chapters of this book for their contributions. We are also very grateful to the reviewers, whose very consistent reviewing of manuscripts was of great help in improving the quality of many chapters of this book. We also owe our gratitude to the publisher of this book for their willingness to deal with the chapters of this book.

Cairo, Egypt   Prof. Aboul Ella Hassanien
                                                                                            Prof. Ashraf Darwish

Contents

Artificial Intelligence in Climate Change Applications

Artificial Intelligence for Predicting Floods: A Climatic Change Phenomenon

Mohamed Torky, Ibrahim Gad, Ashraf Darwish, and Aboul Ella Hassanien

1 Introduction

Climate change is fast becoming one of the biggest and most significant challenges at this time. Many countries in the world are now witnessing radical changes in the features of their climate features. For instance, rainfall styles are shifting, temperatures are increasing, wildfires are starting to occur more frequently, snow and glaciers are melting, the global mean sea level is rising, and floods are becoming more and more dangerous to human life [1]. Climate changes we face confirm that the recent strategies of the world to manipulate them are not up to the challenge we face. For example, these policies will lead to a global temperature rise of $+ 3$ °C. This puts our planet at real risk from more severe climate change effects before the end of the century [2]. Climate change is the large-scale, long-term shift in weather patterns, in particular, due to global warming phenomena. For instance, to make Earth's temperature below the $+ 2$ °C target, CO_2 emissions must be decreased and be close to zero by 2070, with total gas emissions turning negative thereafter [2].

M. Torky (✉)
Faculty of Artificial Intelligence, Egyptian Russian University, Badr City, Egypt
e-mail: mtorky86@gmail.com

I. Gad
Faculty of Science, Tanta University, Tanta, Egypt
e-mail: ibrahim.gad@science.tanta.edu.eg

A. Darwish
Faculty of Science, Helwan University, Cairo, Egypt
e-mail: ashraf.darwish.eg@ieee.org

A. E. Hassanien
Faculty of Computers and AI, Cairo University, Giza, Egypt
e-mail: aboitcairo@cu.edu.eg

M. Torky · I. Gad · A. Darwish · A. E. Hassanien
Scientific Research Group in Egypt (SRGE), Giza, Egypt

© The Author(s), under exclusive license to Springer Nature Switzerland AG 2023
A. E. Hassanien and A. Darwish (eds.), *The Power of Data: Driving Climate Change with Data Science and Artificial Intelligence Innovations*, Studies in Big Data 118,
https://doi.org/10.1007/978-3-031-22456-0_1

On the other hand, flooding is another climatic phenomenon that is possible to occur due to the change in water cycles on our planet. So, the important question is: How does climate change affect floods? Since climate changes occur due to the burning of oil, coal, and gas, this leads to that the weather will change to become warmer, and wetter. You can simulate this scenario by putting some water in a pan and heating it. Increasing temperatures make the water boils, and the steam raises and makes it more moisture, this leads to the constitution of dense clouds that cause heavy rainfalls and occur flooding. There are five climatic causes of flooding: heavy rains, storms, overflowing rivers, lack of vegetation, and melting snow and ice [3]. A more moisture atmosphere can lead to more intense downpours in short times; this can raise the risk of sudden and un-expecting flooding. Moreover, increasing the atmosphere temperature means there is more heat for climate systems that cause heavy rainfalls. Floods shaped the greatest proportion of economic destruction from the unstable climate in Australia over the last 10 years, followed by droughts, and tropical. If these levels of carbon emissions continue without drastic solutions, the Australian economy will lose $40 billion per year by 2060 due to floods [4]. Floods not only have a serious impact on the global economy, and people's life and, but also on our beloved animals, valuable livestock, and wildlife [5]. Hurricane strength is also forecasted to rise as the weather continues to warm. Intense storms can lead to a greater risk of coastal flooding from storm attacks, a danger that will be further growth by sea level rise. Hurricanes occurrences have also serious impacts on moving floods to drink water systems and wastewater services, which can damage treatment procedures or water distribution systems and all infrastructure lines [6].

Flood events are also related to changing rain patterns on snow. In some cases, the downpour-on-snow events occur while the soil is still partially congealed. The frozen soil, which is already saturated, cannot absorb additional water. Therefore, the more rainfall the more snowmelt runoff. This leads to creating many flood rivers with a huge amount of flowing water. This scenario of downpour, frozen soils, and snowmelt was the major reason for Midwest flooding in March 2019 that caused over US$12 billion in damage. In addition, In June 2022, a huge flooding event hit the western mountains series in the Western U.S, where snowmelt combined storms and rainfall dumped up to five inches of rain over three days in Yellowstone National Park and surrounding places, rapidly melting snowpack. Huge amounts of flowing water shaped a dystrophic flood that damaged utilities, and roads, and forced more than 10,000 people to evacuate [7].

Warmer air and wildfires are other reasons for global warming and heavy rains, which generate complex changes in our environmental ecosystems and increase the potential of flood events in various geographic areas on our planet.

Green technology (or environmental technology/green tech.) may be a radical solution to many climate change challenges [8]. Green tech. refers to using science and technology to mitigate the negative environmental impact of human activity and protect the world's natural resources. Although this technology is relatively young, it has fast become a great interest among scientists, engineers, and politicians due to increasing awareness about the impacts of climate change and the exhaustion of environmental resources. The green tech involves four solutions to mitigate climate change and minimize the global warming that increases the potential of flood events [9]:

1. *Low-carbon emission*: It is estimated that our classical homes and building account for 38% of global warming gas emissions around the world [10]. On the other hand, low-carbon houses are designed to require minimal cooling and heating and produce very little waste and pollution. This certainly minimizes the global warming that causes heavy rains and floods.
2. *Global warming reversing*: This process is based on the idea of the possibility of storing and capturing carbon by pulling it from the atmosphere and using it to make synthetic fuel. This solution will help in minimizing atmosphere temperature and global warming. Therefore, the chances of precipitation causing floods are diminishing. Although global warming reversing is still a more expensive task, some studies expect that the costs of direct air carbon storage could come down by a multiple of six, making it a much more achievable process [11].
3. *Renewable energy storage*: storing clean energy such as solar energy for a long time will minimize the dependence on fossil fuels. Burning fossil fuels launch large amounts of CO_2, a greenhouse gas, into the air. Greenhouse gases trap heat in our atmosphere, causing global warming which is a major motivation for rainfall and occurring flooding.
4. *Green Hydrogen Production*: Battery-powered electric vehicles are a common type of electric car, but there's another type of electrical vehicle, which is called a *fuel cell electric vehicle*. These vehicles don't run on batteries, they run on green hydrogen. The main advantage of fuel cell electric vehicles (i.e. Hydrogen-based) is that they don't produce any harmful emissions, and work more efficiently than burning-powered vehicles, and. experts forecast that by 2050, green hydrogen will eventually run over 420 million vehicles and more than 20% of passenger ships [12]. Hydrogen fuel is still expensive, but hydrogen-powered vehicles are already used in many countries all over the world and could become crucial for cutting transport emissions and global warming from fossil fuels-based vehicles.

Artificial Intelligence (AI) is an important technology that can assist in using green technology for mitigating many climates change challenges. Experts and research community and experts have already started focusing on climate informatics with AI models. AI models have a greater potential to predict, assess, classify, recognize, and mitigate climate change hazards with the efficient use of sensing devices, big environmental data patterns, and learning algorithms [13–15]. AI has many diverse applications in climate change areas. For example, AI-based prediction models and decision support systems (DSS) for weather prediction, AI-based machine vision applications in climate informatics, AI-based assessment models to promote eco-friendly energy production and consumption, AI-based prediction models for earth hazard management, AI-based assessment models for reducing the impacts of global warming, Deep learning models for earth informatics and sustainable earth surveillance, AI-based assessment models to minimize carbon footprints for a sustainable environment [16].

Flood prediction is another application of artificial intelligence in climate change area [17]. In flood prediction, predicting flood hazard variables using physical methods can involve a chain of hydraulic and hydrologic models that describe the physical features of floods. Although such models provide flood understanding, they often require high computational and data details. Therefore, Machine-learning techniques can be used as an alternative solution to flood prediction issues. They have the potential to enhance accuracy as well as decrease calculating time and model development cost. This chapter is designed to discuss how AI can be utilized for developing machine/deep learning models to predict flood events. Moreover, the chapter introduces a proposed AI model to add a novel solution to this issue.

The remaining sections of this chapter can be organized as follows. Section 2 presents background about artificial intelligence and flood prediction issue. Section 3 discusses the proposed AI model for predicting flood events. Section 4 presents the experimental results. Section 5 concludes this work.

2 Artificial Intelligence Approaches and Flood Prediction

The recent classical approaches used for flood prediction do not provide accurate forecasting results because of the shortage of sufficient data. Usually, the inaccurate prediction results of those physical approaches lead to catastrophic results as the people cannot timely make the optimal evacuation decisions and put their lives in big danger. One of these traditional approaches is Hydrology [18]. Hydrology is a physical forecasting method that studies the linear relationship between rainfall falling-depth in each time and the resulting water runoff. The sum of this incremental function is then used to estimate the flood flow hydrograph. Another physical method to forecast floods is based on Hydraulic principles [19]. In this way, flood prediction is estimated by studying the relationship between the rainfall pressure on water surfaces and the generated equal forces that push water in all directions.

Figure 1 illustrates how meteorological and hydrological systems are working together for forecasting floods. Note that, all figure's links and rectangles may be required or used. However, the fuzzy data of hydro-meteorological variables such as downpour, water level, flow, and rainfall also result in fuzzy and inaccurate prediction results.

Artificial Intelligence models can do marvels when utilized to anticipate future events such as floods. AI models can mimic non-linear behaviors of a given phenomenon and produce more precise- forecasting results. AI algorithms can predict flood events before they occur, thanks to huge volumes of high-quality datasets. AI-based flood prediction models have some strengths over classical hydrological and hydraulic modeling: (1) Ai-models can provide prediction results and warnings in milliseconds. (2) They work on real observed data and did not consider any hypothetical observations or data structures. (3) They are self-learning methods that can

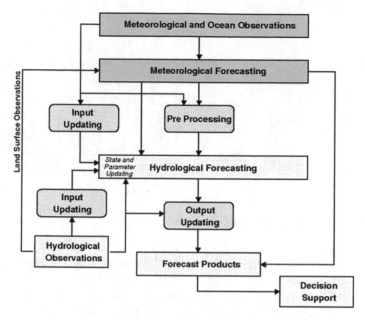

Fig. 1 Meteorological and hydrological flood prediction systems

improve prediction results with time and data. Googles experts believe that AI-flood prediction models can assist in producing flood warnings in a very short time compared to the physical methods and hence can avoid 30 to 50% damage [20].

2.1 How Does AI-Models Work to Forecast Floods

Data is the major requirement for applying AI models. Therefore, in flood prediction issues, sensor devices and satellites are excellent tools for gathering various climatic data patterns from various positions. Based on the collected data, AI algorithms can analyze vectors of data features (e.g. Temperature, rainfall, pressure, wind speed, direction, see and oceans' status, etc.) of this massive information and provide anticipation results about the time and place of flood events.

The methodology of AI-models starts with training (or learning) the used AI-model with the collected dataset (e.g. Climatic data, past rainfall data, temperature, pressure, wind, etc.). Using precise real-world data gathered from governmental and meteorological agencies, the AI- models provide the required flood prediction results. Figure 2 illustrates the AI methodology for flood prediction. The meteorological dataset is input to an AI system as *training data*, which is used to learn a given AI model how to predict flood events based on an extracted vector of data features. After the learning process is terminated, an AI model is then tested with different data patterns (i.e. *testing data*) to verify its accuracy in predicting flood events. The

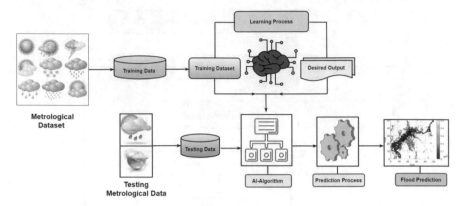

Fig. 2 AI methodology for predicting flood events

learning process in AI is the most significant step with an AI prediction model. The more data volumes the more the learning accuracy of a given AI model.

There are three common learning processes in AI: (1) *supervised learning* is machine learning that requires labeled datasets to train AI algorithms to classify data or predict outputs. (2) *unsupervised learning* is another kind of machine learning in which an AI model is provided with unlabeled and unstructured data. The main objective of this approach is to find structure/patterns in the input data; therefore this learning scheme is called self-learning as an AI algorithm can detect the hidden patterns in unlabeled datasets. (3) *Reinforcement learning* is another learning methodology that reacts to a dynamic environment in which it must achieve a certain goal without any labeled data that can help it to achieve this goal. Therefore, learning a model is based on transactions with a specific domain.

A novel AI-models that can be used for predicting flood events can be developed based on one of the previous learning methods or a hybrid of two of them. Regarding AI prediction models, by specifying only 2–3 features of a data pattern in the historical datasets (e.g. past flood events), the AI prediction model can forecast not only the time and position of flooding but also the seriousness of their occurrence.

Predicting flooding is hypothesized to achieve two goals, predicting flood water level, and predicting flood areas. Therefore, AI can employ two models for achieving these goals respectively, the hydrological Model, which is used to help scientists to predict water levels in rivers, and the invasion model, which determines locations most likely to be damaged by floods. This technique can assist in avoiding damage and mortalities. Moreover, gathered data patterns from social media such as Facebook, Twitter, and other smartphone apps can also help in tracking flooding. Social media-based data holds various data patterns, images, and videos that can help in determining flood locations by AI. Such schemes can monitor, and predict the damage caused by floods [21].

2.2 Machine and Deep Learning for Flood Forecasting

Artificial intelligence could help in developing warning and forecasting systems for forecasting flood events and resulting damage based on machine and deep learning algorithms [22]. Using machine-learning algorithms, floods can be predicted based on structured data, which is used to learn a machine-learning algorithm how to recognize key data features used to predict flood events. Machine learning algorithms can be categorized into three kinds of algorithms, supervised, unsupervised, and semi-supervised algorithms as clarified previously. Deep learning is a subset of machine learning that can perform more complex problems than machine learning based on artificial neural networks (ANN) designs. Therefore, Deep learning algorithms can predict floods based on big and unstructured data volumes in ways machine learning algorithms cannot easily do. Besides performing flood predictions, machine and deep learning algorithms can also monitor and map out/plot flood events while making decision support systems.

Designing machine and deep learning models for flood prediction specifically requires historical data of past flood events and real-time data of rainfall and other meteorological information. The sources of required real-time data involve multi-sensory devices such as radars and satellites, which can monitor the climate in a wide variety of locations. The best machine and deep learning models that can produce optimal results for flood prediction involve the following techniques:

(1) **Artificial Neural Networks (ANNs)**: The neuron-layers design of ANNs makes them very dynamic, adaptable, and effective in learning to model complex flood predictions. These models can assist in flood forecasting floods with high fault tolerance and accurate approximation results. ANN algorithms can be fed with distributed rainfall data from various rain sensors fixed in multiple mountainous regions, then utilize this data to forecast the river water level in valleys between mountain regions. The flow of this river and its water level in these valley regions a few minutes before the flood occurs depict the river's behavior subject to sensor disruption [23]. This real-time flood prediction of an ANN model can be extremely helpful for timely warning and calamity prevention since it can give precise flood prediction within a few seconds. Therefore, ANNs are counted as one of the most dependable AI techniques for designing black-box models of complex and nonlinear relationships between rain flow and flood. As well as river flow and drainage prediction. Figure 3 depicts how ANNs design can be used for flood prediction. The learning process is powered with weights adjusted on neuron links until achieving the best accuracy of flood prediction results. The advance in using ANN models results in the development of deep neural network models able to enhance flood prediction results driven by radar observations and Image recognition based on integrating self-learning functions with neural network models. The highly reliable and precise flood prediction has been achieved by integrating hydrology and river engineering data to develop advanced deep learning models able to predict flood events accurately [24].

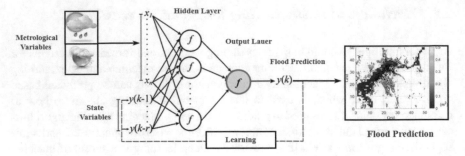

Fig. 3 Flood prediction using ANN model

(2) ***Neuro-Fuzzy Algorithms***: These types of algorithms are defined as mathematical models that employ neural networks and fuzzy logic rules in developing
novel machine learning flood prediction models. In this approach, flood knowledge can be processed using Neuro-Fuzzy Inference System (ANFIS) [25]. The
learning process of this approach is based on an approximation-mathematical
function in developing a nonlinear model for predicting hydrological events
like floods. ANFIS is a popular and efficient flood forecasting approach due to
its fast programming, accurate learning, and powerful abilities for standardization. The ANFIS methodology involves five layers of calculations, input layer,
fuzzification layer, inferences layer, defuzzification layer, and output layer as
depicted in Fig. 4. The Input metrological variables are to be fuzzified in layer 2.
The inference process and fuzzy rules are executed in layer 3. Variables defuzzification of each corresponding rule is carried out in layer 4, and finally, in layer
5, the final flood prediction results are output.

(3) ***Statistical Learning***
    Statistical learning models are popular data-driven techniques, which have
become increasingly used for event regression analysis. These types of models

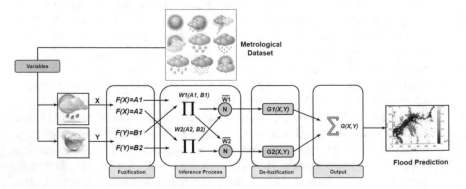

Fig. 4 Flood prediction using ANFIS approach

gave the ability mimic nonlinear hydrological functions. Support Vector Regression (SVR) [26] and Support vector machine (SVM) [27] models are powerful supervised learning models, and their methodology is based on statistical learning theory and the operational risk reduction rule. This makes these models effective techniques to predict flash flood events at different lead times and can serve as an alternative solution to ANNs. The statistical learning–based models can be utilized in various flood prediction cases such as precipitation, extreme rainfall, rainfall-runoff, stream flow, reservoir inflow, soil and weather moisture, and flood quantities and time series. Therefore, SVR and SVM both give promising results, better performance, and remarkably standardized methodologies as compared to ANN.

2.3   AI Systems for Flood Forecasting

Some large companies specializing in information technology and artificial intelligence have begun to develop complete systems and applications that can be greatly relied upon in predicting flood events, warning, and evaluating the risks resulting from the expected occurrence of the flood. These applications can also serve as guidance and recommendation systems for citizens and decision-makers to take the necessary measures to avoid the destruction caused by flood occurrences. In this subsection, we briefly highlight two projects developed as mature systems used in predicting flood events. One of these systems is developed by *Google*, and the other one is developed by *Piccard.AI* organization.

A.   *Google's AI-powered Flood Prediction System*
   Google, the giant of technology has launched an AI-based system in 2017 for predicting river flood events in south Asia [28]. China, India, and Bangladesh are among the regions that are greatly hit by floods and natural disasters with huge damage to infrastructure and loss of lives. Therefore, Google's AI experts developed reliable and scalable flood prediction models depending on machine learning technology. The main idea of Gogol's flood prediction system is based on identifying a river's probability to make flooding based on satellite images and river trajectory. Moreover, Google's system is scalable to involve a set of hydrological variables such as water level, and precipitation to enhance prediction accuracy and extend the lead-time of flood prediction. The pilot version of this system proved that it was able to predict river floods with an accuracy of 75%. Another important advantage of Google's flood prediction system is it can deliver precise real-time flood information and warns the citizens through Google search engine and Google Maps to take defensive precautions. After applying several pilot versions of this system in various regions, Google announced that its flood prediction system issued more than 100 million warning notifications, saving the lives of 360 million citizens in the areas most likely to be attacked by floods. Moreover, the system can launch flood alerts before flood emergence with a lead-time of 72 hours. Although the pilot versions of Google's flood system

proved satisfied with prediction results, it is limited to only river flooding. Therefore, Gogol's AI experts are recently working to update this system to be able to treat not only river floods but also "flash floods", and not only used in the regions covered by Internet area but also in poor regions that are not covered with internet services.

B. *Piccard's AI-based Flash Flood Warning and Prediction System*
Piccard's AI is a software association, which develops AI software systems for prediction, optimization, and automation problems. It has developed a flash flood prediction system able to help residents to avoid any socioeconomic damages. The main idea of this system is based on forecasting rainfall by using IoT sensors, and then, creating alerts before even the downpour occurs. The Piccard's AI system consists of four subsystems: (1) Water level surveillance devices fixed at various hotspot locations. (2) Set of machine learning techniques that can learn from climate data to forecast floods at those hotspot locations before flood occurrence within 1–4 hours. (3) A cloud-based mobile/desktop dashboard for analyzing and presenting flood behavior and its risks. (4) Email/SMS warning system of predicted floods to alert citizens in all the surrounding areas by the predicted flood [29]. The main advantage of this system is its ability to ideally analyze flood behaviors that cannot be addressed with drainage upgrades. By alerting citizens who are at risk, proactively locking roads and, authorities can minimize their damage and reduce the potential risks of flooding.

3 Proposed Machine Learning-Based Flood Prediction Model

In this section, a flood prediction model is proposed, which is based on utilizing machine learning and optimization algorithms. The proposed model consists of three phases: (1) flood data Pre-processing phase, and (2) feature selection based on the Pigeon swarm optimization (PSO) algorithm [30]. (3) Flood prediction based on machine learning phase. Figure 5 illustrates the three phases of the proposed model.

3.1 The Pre-processing Phase

One of the most known stages while developing machine-learning algorithms for data classification is cleaning the used data from any duplication, incompleteness, redundancy, etc. Without cleaning the train data, the classification models will produce accurate and biased results. One of the most popular automated methods that can do the data cleaning phase is SMOTE, which stands for "Synthetic Minority Oversampling Technique". This technique can automatically generate newly cleaned data using the old data. Table 1 shows a sample of the used flood dataset collected from

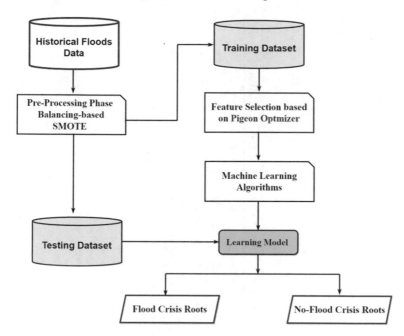

Fig. 5 Proposed machine learning-based flood prediction model

Kerala, India that describes the past floods from 1900 to 2020 [31]. The dataset consists of 120 samples of floods and 14 features. Table 2 shows counts of data labels before and after applying oversampling using SMOTE approach. Figures 1 and 2 shows the scatter plot that visualizes dataset classes before and after applying the SMOTE approach (Figs. 6 and 7).

3.2 Feature Selection Based on the PIO Phase

Pigeon-Inspired Optimization (PIO) is a novel swarm intelligence algorithm inspired by the homing behavior of pigeons [30]. The optimization methodology of the PIO algorithm can be utilized for finding the best solution to a given problem. In flood prediction problems, PIO can be used for finding the optimal reduct of features of the dataset that contribute to obtaining the most precise prediction results. This goal can be achieved by a fitness function to measure the validity of solutions and select the best features of an optimal solution. The used fitness function depends on the total number of features (TF), the performance of the classification model (accuracy), the reduct length, and three weight constants that contribute to the significance of prediction accuray, w_1, subscript and w_2, and w_3, where, $w_1 + w_2 + w_3 = 1$. Equation (1) defines the PIO-fitness function, while Eq. (2) defines the reduct weight that depends on the importance of the feature within the reduct. In our problem, we hypothesized

Table 1 Dataset sample

Subdivision	Year	JAN	FEB	MAR	APR	MAY	JUN	JUL	AUG	SEP	OCT	NOV	DEC	ANNUAL RAINFALL	Floods
Kerala	1901	28.7	44.7	51.6	160	174.7	824.6	743	357.5	197.7	266.9	350.8	48.4	3248.6	Yes
Kerala	1902	6.7	2.6	57.3	83.9	134.5	390.9	1205	315.8	491.6	358.4	158.3	121.5	3326.6	Yes
Kerala	1903	3.2	18.6	3.1	83.6	249.7	558.6	1022.5	420.2	341.8	354.1	157	59	3271.2	Yes
Kerala	1904	23.7	3	32.2	71.5	235.7	1098.2	725.5	351.8	222.7	328.1	33.9	3.3	3129.7	Yes
Kerala	1905	1.2	22.3	9.4	105.9	263.3	850.2	520.5	293.6	217.2	383.5	74.4	0.2	2741.6	No
Kerala	1906	26.7	7.4	9.9	59.4	160.8	414.9	954.2	442.8	131.2	251.7	163.1	86	2708	No

Table 2 The count of data labels before and after applying SMOTE technique

Class label	Before SMOTE	After SMOTE (Oversampling)
Flood crisis	60	60
No-flood crisis	58	60

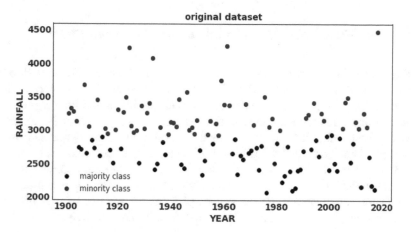

Fig. 6 The distribution of dataset classes **before** applying the SMOTE approach

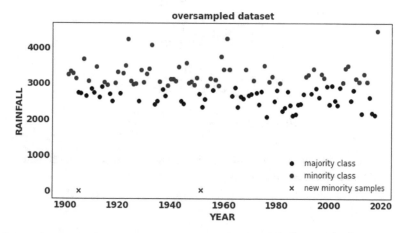

Fig. 7 The distribution of dataset classes **after** applying the SMOTE approach

that $w_1 = 0.70$, $w_2 = 0.22$, and $w_3 = 0.08$.

$$Fitness\ function = w_1 * accuracy + w_2 * F_1 + w_3 * \left(1 - \frac{SF}{TF}\right) \quad (1)$$

$$\text{Reduct weight} = \sum_i \text{importance}(fi_i) * \frac{TF - SF}{TF}, \text{ such that importance}(fi_i)$$
$$> \text{threshold} \tag{2}$$

3.3   Flood Prediction Based on Machine Learning Phase

In this phase, two learning models have been applied, called Gradient Boosting (GB) [32], and Random Forest (RF) [33], but with two different scenarios. In the first scenario, the GB algorithm is combined with the PIO algorithm for predicting floods, while, in the second scenario, the process of flood prediction is completely solved by the RF algorithm without using the PIO algorithm. The two scenarios have been applied and compared to identify the features of the flood crisis and predict its occurrence. The learning process is performed before and after applying SMOTE technique in the trained dataset to validate the efficiency of the proposed model. The performance is validated using the known performance metrics such as precision, recall, accuracy, and F1-score as in Eqs. (3, 4, 5, and 6) respectively.

$$\text{Precision} = \frac{TP}{TP + FP} \tag{3}$$

$$\text{Recall} = \frac{TP}{TP + FN} \tag{4}$$

$$\text{Accuracy} = \frac{TP + TN}{TP + TN + FP + FN} \tag{5}$$

$$\text{F1 - Score} = \frac{2(\text{Precision*Recall})}{\text{Precision} + \text{Recall}} \tag{6}$$

where, TP, TN, FN, and FP are truly positive, true Negative, False Negative, and False Positive numbers respectively.

4   Experimental Results

In this section, the effectiveness of the proposed flood prediction model is evaluated using a Kerala, India dataset. The proposed model has been developed in Python and the *scikit-learn* software packages. The testing method is carried out on a *Google Colab cloud* equipped with a CPU running at 2.6 GHz and 32 GB of RAM. Two scenarios of experiments have been carried out, and the obtained results can be detailed as follows.

4.1 Flood Prediction Results Using PIO and GB Algorithms

In this scenario, a set of preparation steps are carried out for the dataset before starting the experiment and used as an input to the selected machine learning algorithms.

The preparation steps are carried out for the dataset before it can be utilized as an input for machine learning algorithms. The dataset consists of a set of attributes that describe the nature of the data. A lot of combinations are evaluated to select the most significant attributes based on their impacts on classification performance. The suggested PIO reduct technique is tested on the flood dataset, containing the monthly and yearly. Additionally, the Gradient Boosting Classifier (GBC) is used to evaluate the performance of this technique. The suggested PIO is evaluated using accuracy and F1-score, appropriate for this classification challenge.

Table 3 provides a summary of the data. The dataset has various features and instances ranging from 1 to 14 and 1 to 120. The total number of states is 1, with dimension data (120, 13). Therefore, 80% of the rows were selected for the training set, and 20% were used for testing purposes.

Many attributes in this dataset describe the flood problem. To develop a robust classification model, all characteristics are not necessary. A subset of these features can be selected to improve the accuracy and the F1-score values. The updated dataset is utilized for training the selected ML models once the optimal reduct has been identified using the PIO feature selection technique.

Figure 8 shows the confusion matrix and stability score for the reduction, and it is clear that {FEB, JUN, SEP, OCT, NOV, and ANNUAL RAINFALL} can be considered as a common features that improved the classification performance. The reduction that consists of these features has a high stability value of 0.96, which means that it is the most significant variable that can accurately describe the flood crisis. Also, the confusion matrix is considered an effective measure since the reduct that contains these features has a correct NO value of 4 (100%) and a correct YES value of 8 (100%) as shown in Fig. 9.

Similarly, Fig. 10 is another example to show the results of different reducts. From the figure, it is clear that {JAN, JUN, JUL, AUG, and DEC} can be considered as common features that improved the classification performance. The reduction that consists of these features has a high stability value of 0.971, and the accuracy value for training data is 100%. Where the reduct that consists of these features has an accuracy value of 100% for testing data. Also, a confusion matrix is included as an

Table 3 The description of the dataset		Value
	Country	Kerala, India
	Period	1900–2020
	Samples	120
	Attributes	14
	Classes	2 (Flood/No-flood)

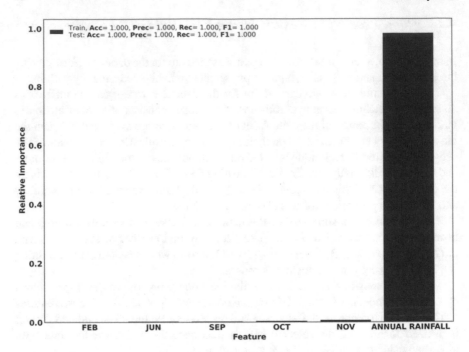

Fig. 8 The performance metrics for the reduct {(FEB, JUN, SEP, OCT, NOV, and ANNUAL RAINFALL)}

Fig. 9 The confusion matrix of the proposed model for the reduct {(FEB, JUN, SEP, OCT, NOV, and ANNUAL RAINFALL)}

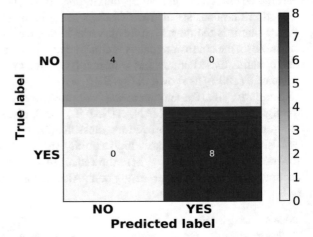

effective measure. The reduction that contains these features has a correct NO value of 4 (100%) and a correct YES value of 8 (100%).

Due to class imbalance in test data and accuracy, F1 metrics are used to train and test predictions using scikit-learn's library. Table 4 presents the feature selection results of the suggested PIO method on the flood dataset. For 10 iterations, the

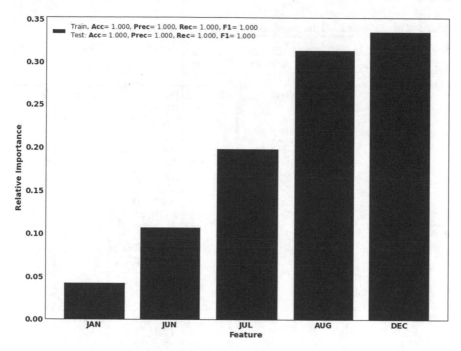

Fig. 10 The reduct Stability 0.971 for the reduct {(JAN, JUN, JUL, AUG, and DEC)}

proposed algorithm has chosen 2 to 13 features out of the available 14 features. The obtained weight is also provided, the lower the F1, the worst performance. For example, the proposed method has selected 7 features with a weight of 0.4305 at the second iteration. From Table 4, it is found that the PIO method is performing well.

Figure 11 provides the frequencies of each reduce the length of 10 iterations for the proposed method. The PIO method chose the highest reduct size with 12 features and the least number of features necessary, which was three out of the total of 14 features. According to this figure, the reduct size 7 is the most common one; it appears 29 times. Similarly, the reduct sizes 11, 3, and 12 are produced a subset of features with the frequencies of 4, 3, and 1, respectively.

Table 5 shows the fitness value, accuracy, and F1-score of the proposed model for each iteration. The fitness function, the global solution Xgb and the local solution Xpg for each iteration are presented in Table (5). Moreover, in the whole iteration, it is noticed that the PIO produced better results with the maximum accuracy of 100% and F1 and fitness function values of 100% and 100%, respectively. From the available 14 features, PIO selected an average of 3 features. The results show that the PIO method achieves an acceptable level of performance throughout all the iterations.

Table 4 The performance of the proposed model and the weight for each iteration

Iteration	Train		Test		Reduct	Weight	Reduct size
	Accuracy (%)	F1-Score (%)	Accuracy (%)	F1-Score (%)			
0	100	100	100	100	['YEAR', 'FEB', 'MAR', 'MAY', 'JUN', 'JUL', 'AUG', 'OCT', 'NOV', 'ANNUAL RAINFALL']	0.280852	10
2	100	100	100	100	['YEAR', 'FEB', 'APR', 'JUN', 'JUL', 'AUG', 'NOV', 'DEC']	0.325611	8
2	100	100	100	100	['YEAR', 'FEB', 'APR', 'JUN', 'JUL', 'AUG', 'NOV']	0.4305	7
2	100	100	100	100	['YEAR', 'JAN', 'FEB', 'MAY', 'OCT', 'NOV', 'ANNUAL RAINFALL']	0.493	7
3	100	100	100	100	['YEAR', 'FEB', 'MAR', 'APR', 'JUN', 'JUL', 'AUG', 'SEP', 'NOV', 'DEC']	0.236236	10
6	100	100	100	100	['JAN', 'FEB', 'MAY', 'AUG', 'SEP', 'OCT', 'NOV', 'ANNUAL RAINFALL']	0.422565	8

(continued)

Table 4 (continued)

Iteration	Train		Test		Reduct	Weight	Reduct size
	Accuracy (%)	F1-Score (%)	Accuracy (%)	F1-Score (%)			
7	100	100	100	100	['YEAR', 'FEB', 'JUL', 'AUG', 'NOV', 'DEC', 'ANNUAL RAINFALL']	0.493	7

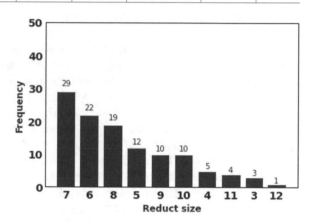

Fig. 11 The frequencies of each reduct

Table 5 The performance of the proposed model for each iteration

	Xgb			Xpg		
Fitness value	Accuracy	F1-score	Reduct	Accuracy	F1-score	Reduct
0.982857	1.000	1.000	[5, 6, 11]	1.000	1.000	[0, 1, 2, 4, 7, 10]
0.982857	1.000	1.000	[5, 6, 11]	1.000	1.000	[0, 2, 8, 11]
0.982857	1.000	1.000	[5, 6, 11]	1.000	1.000	[0, 3, 4, 6, 8, 9]
0.982857	1.000	1.000	[5, 6, 11]	1.000	1.000	[4, 7, 8, 9, 10, 12]
0.982857	1.000	1.000	[5, 6, 11]	1.000	1.000	[1, 6, 8, 12]
0.982857	1.000	1.000	[5, 6, 11]	1.000	1.000	[0, 3, 10, 12]
0.988571	1.000	1.000	[9, 11]	1.000	1.000	[9, 11]
0.988571	1.000	1.000	[9, 11]	1.000	1.000	[7, 8, 13]
0.988571	1.000	1.000	[9, 11]	1.000	1.000	[1, 4, 9,13]
0.988571	1.000	1.000	[9, 11]	1.000	1.000	[4, 6, 10]

Table 6 The performance results for the random forest model on the flood dataset

Model	Train				Test			
	Accuracy	Precision	Recall	F1-score	Accuracy	Precision	Recall	F1-score
Random Forest	1.0	1.0	1.0	1.0	1.0	1.0	1.0	1.0

4.2 Flood Prediction Results Validation Using Random Forest (RF) Algorithm

The Random Forest model has been applied and tested on the obtained features that optimized by the PIO algorithm. Table 6 summarizes training and testing results of RF algorithm. In the reduction step, the PIO method based on the Gradient Boosting model selected an average of 7 features. Then, after the feature selection step, the Random Forest model attained better performance in both of training and testing steps in terms of accuracy. Table 6 shows that the Random Forest method achieves in training step accuracy of 100%, F1 of 100%, significantly equal to the testing step.

Figure 12 shows the confusion matrix for the RF model. The reduct that contain {"YEAR", "JAN", "FEB", "APR", "JUN", "NOV", "ANNUAL RAINFALL"} features has a correct "NO" value of 4 and correct "YES" value of 8. Receiver Operating Characteristic (ROC) curves are important assistants in evaluating the classification models. The ROC curve is a plot of False Positive Rate (FPR) on the x-axis and True Positive Rate (TPR) on the y-axis. The best value of the area under the ROC curve of the test data is 1. Figure 13 shows the ROC curve for the RF model. It is clear that the ROC curve passes through the upper left corner (100% FPR and 100% TPR).

Fig. 12 Confusion matrix of RF model based on the optimized features using PIO algorithm

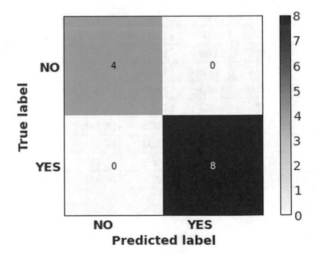

Fig. 13 ROC curve for RF model

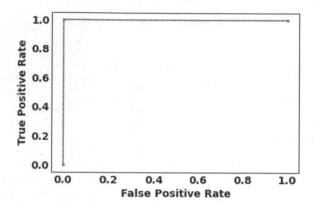

Table 7 shows the performance of Gradient Boosting (GB), and RF models on the dataset after performing feature selection step by PIO algorithm. In this dataset, the PIO method and the Gradient Boosting model chose an average of 7 features and attained the best accuracy of 100%. This results has been confirmed when the Random Forest algorithm has been applied on the optimized features set. The practical results showed that RF achieved an accuracy of 100%, significantly equal to the GB method. This confirms the efficiency of PIO in selecting the best features contribute flood prediction.

Figures 14 and 15 shows equal confusion matrices for the RF, and Gradient Boosting models. This confirms stability and efficiency of PIO for performing features selection. The obtained reduct of best features achieved optimal classification results whatever the used classifier, RF or Gradient Boosting.

Table 7 Comparative performance results for the different machine learning models on the collected dataset

Classifier	PIO		SMOTE			
	Accuracy (%)	Reduct	Accuracy (%)	Precision (%)	Recall (%)	F1-score (%)
Gradient boosting	100	["YEAR, "JAN", "FEB", "APR", "JUN", "NOV", "ANNUAL RAINFALL"]	100	100	100	100
Random forest	100	["YEAR, "JAN", "FEB", "APR", "JUN", "NOV", "ANNUAL RAINFALL"]	100	100	100	100

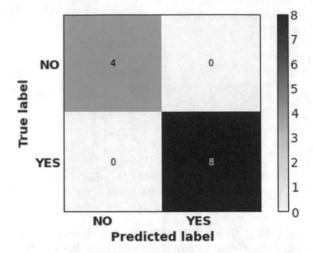

Fig. 14 Confusion matrix for RF model

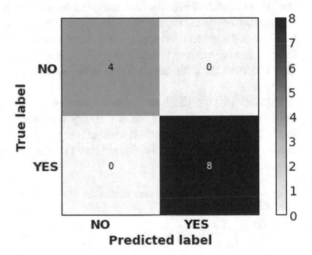

Fig. 15 Confusion matrix for gradient boosting model

5 Conclusion

This chapter has discussed the issue of flood prediction, one of the most important challenges in climate change. This study has identified the majority of using artificial intelligence techniques for forecasting the onset of flooding based on machine learning, deep learning, and neuro-fuzzy algorithms compared to the traditional methodology that depends on hydrology and hydraulic. The results of this investigation indicate that predicting floods based on AI approaches can develop novel smart systems that can save the lives of millions of citizens in the areas most likely to be attacked by floods. Moreover, these systems can launch flood alerts before flood emergence with a lead-time of 72 h. Those strengths facilitate evacuating residents and enable decision makers to make suitable decisions by which damage, losses in lives, money, and infrastructure will reduce to the minimum level. The second major finding was introducing a novel machine learning model that utilized the Pigeons Swarm Inspired Optimization (PIO) algorithm for selecting the optimal reduct of floods features that contribute to predicting the onset of floods precisely. Taken together, these findings confirm the role of artificial intelligence in developing additional machine and deep learning models for predicting natural disasters and climate changes.

References

1. VijayaVenkataRaman, S., Iniyan, S., & Goic, R. (2012). A review of climate change, mitigation, and adaptation. *Renewable and Sustainable Energy Reviews, 16*(1), 878–897.
2. AON. (2018, September). *Climate change challenges.* [online], Available https://www.aon.com/getmedia/8ddb2a56-c1a9-4689-81e6-f3b7c178e57c/Climate-Change-Challenges.aspx. Accessed July 19, 2022.
3. B-Air. (2018, February 27). *What causes floods? Top 8 common causes of flooding.* [Online], Available https://b-air.com/2018/02/common-causes-flooding/. Accessed July 19, 2022.
4. Climate Council. (2022, March 2). *Everything you need to know about floods and climate change.* [Online], Available https://www.climatecouncil.org.au/resources/. Accessed July 19, 2022.
5. Climate Council. (2021, December 17). *How to care animals in a flood.* [Online], Available https://www.climatecouncil.org.au/care-for-animals-flood/. Accessed July 19, 2022.
6. APA. (2022, July 5). *Climate adaptation and storms and flooding.* [online], Available https://www.epa.gov/arc-x/climate-adaptation-and-storms-flooding. Accessed July 19, 2022.
7. The conversation. (2022, July 31). *Climate change is making flooding worse: 3 reasons the world is seeing more record-breaking deluges and flash floods.* [Online], Available https://theconversation.com/. Accessed July 19, 2022.
8. Hossain, M. F. (2017). Green science: Independent building technology to mitigate energy, environment, and climate change. *Renewable and Sustainable Energy Reviews, 73*, 695–705.
9. Normand, A. (2022, March 11). *Everything you need to know about green technology in 2022.* [online], Available https://www.greenly.earth/blog-en/. Accessed July 20, 2022.
10. Climate colab. (2016, August 12). *Low carbon building.* [online], Available https://www.climatecolab.org/contests/2016/buildings/c/proposal/1329602. Accessed July 20, 2021.
11. Breyer, C., Fasihi, M., Bajamundi, C., & Creutzig, F. (2019, September 18). Direct air capture of CO_2: A key technology for ambitious climate change mitigation. *Joule, 3*(9), 2053–2057.
12. Rogers, M. (2019, January 7). *These 9 technological innovations will shape the sustainability agenda in 2019.* [online], Available https://www.mckinsey.com/business-functions/sustainability/our-insights/sustainability-blog/. Accessed July 20, 2022.
13. Leal Filho, W., Wall, T., Mucova, S. A., Nagy, G. J., Balogun, A. L., Luetz, J. M., Ng, A. W., Kovaleva, M., Azam, F. M., Alves, F., & Guevara, Z. (2022, July 1). Deploying artificial intelligence for climate change adaptation. *Technological Forecasting and Social Change, 180*, 121662.
14. Fathi, S., Srinivasan, R. S., Kibert, C. J., Steiner, R. L., & Demirezen, E. (2020, April 16). AI-based campus energy use prediction for assessing the effects of climate change. *Sustainability, 12*(8), 3223.
15. Toniuc, D., & Groza, A. (2017, September 7–9). Climebot: An argumentative agent for climate change. In *Proceedings of 2017 13th IEEE international conference on intelligent computer communication and processing (ICCP)*, (pp. 63–70).

16. Xin, Q., Samikannu, R., & Wei, C. (2022, July 5). Special issue: Artificial intelligence for climate change risk prediction, adaptation, and mitigation, ecological processes. Springer [online], Available https://ecologicalprocesses.springeropen.com/. Accessed July 20, 2022).
17. Mosavi, A., Ozturk, P., & Chau, K. W. (2018). Flood prediction using machine learning models: A literature review. *Water, 10*(11), 1536.
18. Sene, K. (2010). Hydrological forecasting. In S. Kevin (eds) *Hydrometeorology*, (pp. 101–139). Springer [online], Available https://doi.org/10.1007/978-90-481-3403-8_4
19. Nile, B. K. (2018). Effectiveness of hydraulic and hydrologic parameters in assessing storm system flooding. *Advances in Civil Engineering, 2018*, 4639172.
20. Sharma, B. (2021, November 11). *Google's AI-based flood forecasting system is saving lives in India: Here's how*. [online], Available https://www.indiatimes.com/technology/news/. Accessed July 12, 2022.
21. Thinkml. (2022, February 4). *Flood predictions using AI*. [online], Available https://thinkml. ai/flood-predictions-using-ai/. Accessed July 21, 2022.
22. Shinde, P. P., & Shah, S. (2018, August 16–18). A review of machine learning and deep learning applications. In *Proceedings of 2018 fourth international conference on computing communication control and automation (ICCUBEA)*, (pp. 1–6). Pimpri Chinchwad College of Engineering.
23. Fazel, S. A., Mirfenderesk, H., Blumenstein, M., & Tomlinson R. (2014, January 1). Application of a neural network to flood forecasting, an examination of model sensitivity to rainfall assumptions. In *Proceedings of the 7th international congress on environmental modelling and software*, June 15–19, (pp. 1–9).
24. Park, K., Jung, Y., Seong, Y., & Lee, S. (2022). Development of deep learning models to improve the accuracy of water levels time series prediction through multivariate hydrological data. *Water, 14*(3), 469.
25. Sahoo, A., Samantaray, S., Bankuru, S., & Ghose, D. K. (2020). Prediction of flood using adaptive neuro-fuzzy inference systems: A case study. In S. Satapathy, V. Bhateja, J. Mohanty, & S. Udgata, (Eds.), *Smart intelligent computing and applications. Smart innovation, systems and technologies*, (Vol. 159). Springer. https://doi.org/10.1007/978-981-13-9282-5_70
26. Wu, J., Liu, H., Wei, G., Song, T., Zhang, C., & Zhou, H. (2019). Flash flood forecasting using support vector regression model in a small mountainous catchment. *Water, 11*(7), 1327.
27. Yan, J., Jin, J., Chen, F., Yu, G., Yin, H., & Wang, W. (2018). Urban flash flood forecast using support vector machine and numerical simulation. *Journal of Hydroinformatics, 20*(1), 221–231.
28. Jeremy, K. H. (2020, September 1). *Google bolsters its A.I.-enabled flood alerts for India and Bangladesh*. [online], Available https://fortune.com/2020/09/01/google-ai-flood-alerts-india-bangladesh/. Accessed July 24, 2022.
29. Microsoft. *AI flash flood forecasting*. [online], Available https://appsource.microsoft.com/en-us/product/web-apps/piccardai.piccard-flood-forecasting?tab=overview. Accessed July 27, 2022.
30. Duan, H., & Qiu, H. (2019). Advancements in pigeon-inspired optimization and its variants. *Science China Information Sciences, 62*(7), 70201.
31. Thakur, M. (2020). *Flood prediction model*. [online], Available https://www.kaggle.com/code/mukulthakur177/flood-prediction-model/data. Accessed September 1, 2022.
32. Xu, Z., Huang, G., Weinberger, K. Q., & Zheng, A. X. (2014 August, 24–27). Gradient boosted feature selection. In *Proceedings of the 20th ACM SIGKDD international conference on Knowledge discovery and data mining*, (pp. 522–531).
33. Hastie, T., Tibshirani, R., & Friedman, J. (2009). Random forests. In H. Trevor, T. Robert, & F. Jerome (Eds.), *The elements of statistical learning* (pp. 587–604). Springer.

Prediction of Climate Change Impact Based on Air Flight CO$_2$ Emissions Using Machine Learning: Towards Green Air Flights

Heba Askr, Aboul Ella Hssanien, and Ashraf Darwish

1 Introduction

As stated by United Nations' Food and Agriculture Organization (FAO) report from 2022, A major issue facing the entire world is climate change which calls for broad and sector-wide action. Such action must be performed while fully considering international objectives and agreements, like the Sustainable Development Goals for 2030 [1].

Based on sustainable development goal number 'Thirteen which is called 'climate action', all nations on all continents are being impacted by climate change. It is harming people's lives and upsetting national economies. The weather is changing, the ocean is rising, the severity of meteorological occurrences is rising, and action must be taken quickly to save lives to address the climate pandemic [2].

The main reason for climate change is the total greenhouse gas emissions (GHGs). Power imbalances in the atmosphere are brought on by excessive GHG emissions such that the Earth's system receives more energy than it emits, causing global warming. The behavior of the rapidly rising atmosphere is very different, showing important changes in weather patterns that ultimately has an impact on our global

H. Askr (✉)
Faculty of Computers and Artificial Intelligence, University of Sadat City, Sadat, Egypt
e-mail: heba.askr@fcai.usc.edu.eg
URL: http://www.egyptscience.net

A. E. Hssanien
Faculty of Computers and Artificial Intelligence, Cairo University, Cairo, Egypt
URL: http://www.egyptscience.net

A. Darwish
Faculty of Science, Helwan University, Cairo, Egypt
URL: http://www.egyptscience.net

H. Askr · A. E. Hssanien · A. Darwish
Scientific Research Group in Egypt (SRGE), Giza, Egypt

© The Author(s), under exclusive license to Springer Nature Switzerland AG 2023   27
A. E. Hassanien and A. Darwish (eds.), *The Power of Data: Driving Climate Change with Data Science and Artificial Intelligence Innovations*, Studies in Big Data 118,
https://doi.org/10.1007/978-3-031-22456-0_2

environment. These effects will become worse at average global temperature of 1.5 °C and more so at 2.0 °C as stated in the Paris Agreement (United Nations Framework Convention on Climate Change (UNFCCC), 2015). If climate change is not adequately or completely mitigated, by 2100, it's anticipated that the average global temperature would rise to 3 °C, or even higher with considerably more disastrous consequences [3].

The environment is vital for both individuals and businesses since businesses can exploit the environment as a resource. Environmental harm is therefore quite important for businesses as well. There has been a sharp increase in global demand for air travel and it is expected in 2035 there will likely be 7.2 billion aviation travelers worldwide according to the International Air Transport Association (IATA) [4]. On the other hand, air flights have a big climate change impact as per 1000 passenger-kilometers traveled, a flight generates on average 18 times as much CO_2 as a journey by train [5].

ML is a subfield of Artificial Intelligence (AI) that makes it possible for computers to directly learn from examples, data, and experience. ML systems may carry out complex processes by learning from data rather than adhering to pre-programmed rules by enabling computers to accomplish certain jobs intelligently [6]. ML has the potential to be a potent tool for lowering GHG emissions and assisting society in climatic adaptation because ML has the potential to result in significant efficiency gains, enhance public services, aid in assembling facts for decision-making, and direct strategies for future growth.

The green economy is defined by Zhironkin, and Cehlár in [7] as the most recent method of getting and using resources. It embodies many of the accomplishments of Industry 4.0 and is a result of the Fourth Industrial Revolution. The changes in the economy are brought on by the rise of new waste recycling industries, energy generation with no emissions, emission reduction of GHGs, environmentally friendly urbanism, and likewise post-mining. However, the only way to truly transition to a green economy is to ensure that every industry develops sustainably, and that green technology are used in every aspect of production and consumption. In fundamental industries, green manufacturing should be fostered (mining, energy, engineering, chemistry, transport), as well as in high-tech sectors that open new avenues for environmentally conscious modernizing.

Towards a green air flight, green economy, and accordingly a green environment, this paper developed an ML model to predict the climate change impact resulting from the air flight CO_2 emissions on the environment.

This chapter is organized in the following manner. Section 2 gives a preliminary for the Decision Tree and Random Forest classifiers. Section 3 presents the methodology of the proposed model. Section 4 discusses the obtained results and the performance evaluation. Section 5 gives the conclusion of the paper.

2 Preliminary

A decision Tree (DT) is a resource for predictive modelling with a wide range of applications. They can be created using an algorithmic method that can divide the dataset in many ways according to various conditions. DTs are considered the strongest supervised algorithms. Both classification and regression can be accomplished using them. A tree's two primary components are decision nodes, in which the data is divided, and leaves, where we obtained a result [8]. The main idea of the traditional DT algorithm is presented in Fig. 1. A DT is a decision support technique that forms a tree-like structure. It consists of three components: a root node, decision nodes, and leaf nodes. The nodes in the decision tree present features which are employed to predict the result. A DT algorithm divides a training dataset into branches based on the decision node, which is further divided into other branches. This pattern keeps going until a leaf node is attained (the output or the target) and the leaf node cannot be divided further.

Random Forest (RF) is a supervised learning algorithm. This is employed in both regression and classification. But it is primarily employed for classification issues. As we are aware, trees make up a forest, and a stronger forest results from having more trees. Likewise, the RF algorithm generates DTs from data samples, and subsequently obtains each own prediction and then votes to determine which option is the best. It is a collective approach, in the case of large-scale issues, is superior to a single DT where there are a great number of features. This is because it lessens the excessive fitting using the outcome's average [9]. As shown in Fig. 2, A traditional RF algorithm consists of many DTs, and it determines the result based on the predictions of these DTs. In every RF tree, a randomly chosen collection of features is used at the split of the node.

Fig. 1 The main idea of the traditional DT algorithm

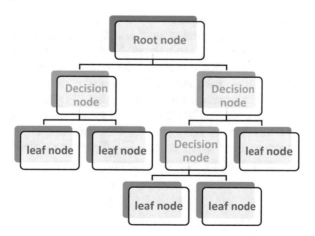

Random Forest Algorithm

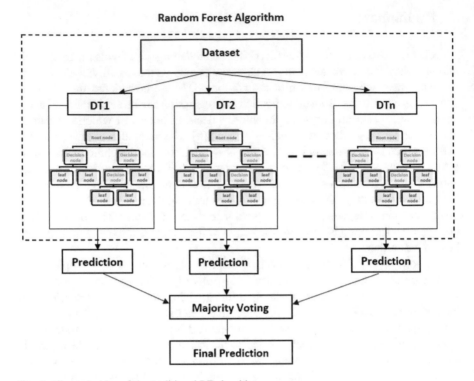

Fig. 2 The main idea of the traditional RF algorithm

3 Methodology

ML is a system's capacity to automatically gather, combine, and then create knowledge from massive amounts of data, and then independently advance the information gained, without having been explicitly coded. ML algorithms can comprehend the produced data being studied, take a model of the pattern, and store it, forecast the future values that the built-in model will produce, and aggressively look for any abnormal phenomenon behavior so that it is possible to take preventative corrective action in advance [10]. Therefore, in this paper, we used two ML classifiers based on a threshold of 100 tonnes for CO_2 emissions to predict if there are CO_2 impacts or influences of the investigated air flights on the environment. The proposed model is described in the next subsections.

3.1   Proposed Model

In this chapter, a binary classification ML model was developed to predict the air flight climate change impact for 186 countries. RF and DT algorithms are individually trained, validated, and tested. The proposed model is presented in Fig. 3. The proposed model is composed of several DTs (which comprise a random forest). Each DT consists of 3 decision levels based on whether the CO$_2$ emissions are greater than a predefined threshold value (100 tones of CO$_2$) or not, whether the flight type is a passenger or not, and whether the country is in one of the top ten countries that provide a huge quantity of CO$_2$ emissions or not. The proposed model has been run two times, the first one using the DT algorithm and the other one using the RF (DT1, DT2, ..., DTn) algorithm.

3.2   Dataset Characteristics and Pre-processing

We collect more than 145,000 data records regarding the CO$_2$ emissions derived from air flights all over the world through the Organization for Economic Cooperation and Development (OECD) [11]. It is a global organization that strives to create better laws for better lives. Its objective is to create laws that promote everyone's prosperity, equality, opportunity, and well-being. It contains more than 60 years of knowledge and understanding that will help better prepare the world of the future through creating worldwide standards based on solid data and resolving a variety of social, economic, and environmental problems. In addition, it offers a distinctive platform and resource centre for data and analysis, shared experiences, exchange of best practices, providing suggestions for governmental policies, as well as defining international standards. From its huge databases, we extract the air flight dataset which we used in this paper. This dataset contains data about the CO$_2$ emissions in tonnes for 186 countries all over the world, associated with flight type (passenger or freight) and flight frequency (annually, monthly, quarterly), travel time (from 2014 to 2022), and sources of CO$_2$ emissions. The description of the data variables used in this paper is shown in Table 1.

A crucial stage in ML is data pre-processing because the value of the data and the knowledge that may be gained from it have a direct impact on how well our model learns; Consequently, we must pre-process our data before incorporating it into our model. As we aimed to predict if the features of each air flight will lead to an impact on climate change, we add a new feature called" climate_change_impact". First, during the data cleaning, there were no empty values relating to the air flight dataset through the 145,502 records and thus did not need imputation.

Second, there are five categorical (nominal) variables which are 'Country', 'Flight Type', 'Frequency', 'Time' and 'Source of CO$_2$ Emissions', and two numerical variables which is 'CO$_2$ Emissions' and 'climate_change_impact'. Therefore, the

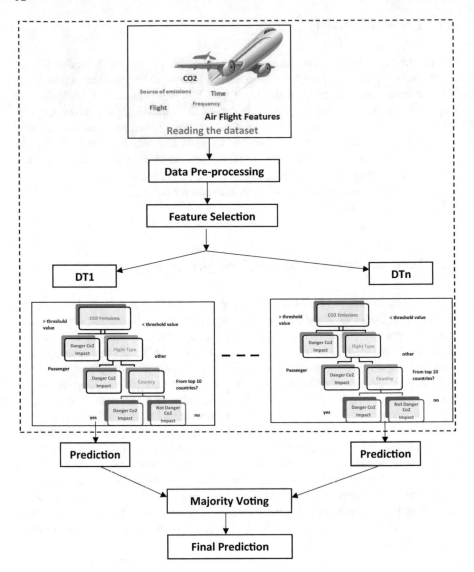

Fig. 3 The proposed model

nominal variables were numerically transformed and coded. Finally, following data
transformation and cleansing, Scaling was done to the variables.

Table 1 Data variables description

Variable	Description
Country	The string variable represents the name of the country
Flight type	The string variable represents the type of the flight (passenger or freight)
Frequency	The string variable represents the flight travel frequency (annually, monthly, or quarterly)
Time	The string variable represents the flight year, a quarter of a year, or a monthly slot
CO_2 emissions	The Numeric (float) variable represents the tonnes of CO_2 emissions resulting from the investigated data record
Source of emissions	The string variable represents the source of CO_2 emissions

3.3 Feature Selection

To create prediction models, feature selection is significant since features have a direct impact on how well models perform [12]. Correlation plays a very important role in finding the dependency among the features of the dataset. The heatmap in Fig. 4 is drawn to visualize the correlation among the variables (features) where darker shades represent a positive correlation. In our dataset, the most important features which affect the "climate_change_impact" are "CO_2_Emmisions"," Flight type", and" Country".

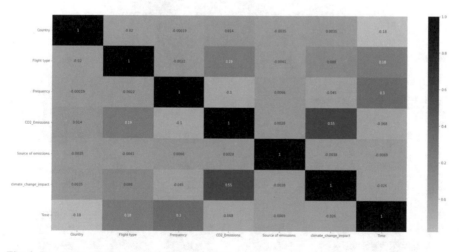

Fig. 4 The heatmap of the investigated dataset

4   Results and Performance Evaluation

Because the accuracy is the most popular performance indicator for classification methods, the performance of each algorithm is evaluated by the accuracy of five folds. The implementation of this chapter is carried out with a binary classification ML model over a time series (2014–2022) dataset to provide the predicted target (have an impact on climate change or not). After excluding the less important features from our dataset by using the heatmap visualization in the previous section, it is also found by the random forest feature importance that the most important feature that affects the target ('climate_change_impact') is the CO_2_Emissions as shown in Fig. 5.

RF and DT models are trained individually on the dataset. Results are produced by using both models individually in our testing set resulting in a 0.99 output accuracy by both models as shown in Fig. 6.

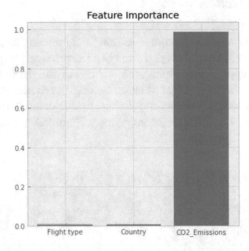

Fig. 5 RF feature importance

	precision	recall	f1-score	support
0	1.00	1.00	1.00	4005
1	1.00	1.00	1.00	25095
accuracy			1.00	29100
macro avg	1.00	1.00	1.00	29100
weighted avg	1.00	1.00	1.00	29100

The accuracy for 5 : 0.9999800757122933

Fig. 6 The accuracy after 5 folds

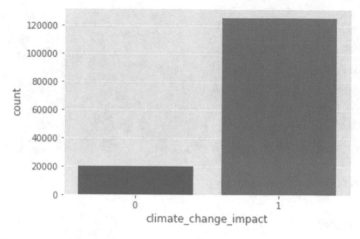

Fig. 7 Climate change impact indicator

As you can see from Fig. 7, among the 186 countries all over the world, the number of the countries that will affect the climate change through the periods 2014 to 2022 because of using air flights has registered a big value as compared to the other countries that have not dangerous climate change impact where" count" represents the number of the countries.

To make the graph more obvious for the readers, we took a snapshot of the top 20 registered CO$_2$ emissions as shown in Fig. 8. It is observed that the largest value of CO$_2$ emissions is registered in 2019 and the smallest value is in 2017. It is also observed that the CO$_2$ emissions in the first quarter of 2022 exceeded the same quarter in 2019, 2020, and 2021 and this is a very important indicator that the CO$_2$ emissions by the end of 2022 may exceed the value registered in 2019. This prediction put a spotlight on the countries which are responsible for these emissions and the disaster of climate change throughout the world.

Figure 9 presents the top 10 countries from 2014 to 2022 which provide the largest values of CO$_2$ Emissions based on their air flights and which have the most impact on the climate change indicator. It is observed that the largest CO$_2$ emissions were provided by the United States, China, United Arab Emirates, United Kingdom, Germany, Japan, France, Korea, Russia, and Canada.

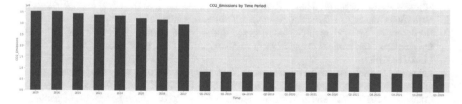

Fig. 8 Top 20 CO$_2$ emissions by time-period

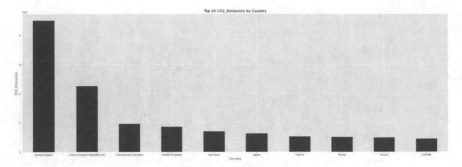

Fig. 9 Top 10 CO$_2$ emissions by country

5 Conclusion

The use of aircraft is what emits GHGs and carbon dioxide (CO$_2$) and is one of the key factors that contribute to global climate change. In this paper, we developed a ML model to predict if there is a dangerous impact of air flights' CO$_2$ emissions on climate change for the investigated 186 countries all over the world from the years 2014 to 2022. The paper aims to spot the light on the countries which deeply affect climate change through their air flights' CO$_2$ emissions. This is done based on a recent and up-to-date website containing 145,502 records to help the approach of transforming to green air flights and green economy in the world and adaptation to the predicted future climate change pandemic. There is a need to calculate the air flight emissions to save the environment from the expected damaged climate change using developed tools such as Zurich Airport's Aircraft Local Emissions Calculator for Airports (ALECA) which is a tool for independently calculating emissions for every source of emissions connected to aviation at an airport and the ACI Aircraft Ground Energy System which is a Simulator (AGES-Simulator) which assesses the advantages on both the environment and the economy of substituting the use of the Auxiliary Power Unit (APU) by AGES by calculating the decrease in fuel use.

References

1. https://www.fao.org/3/ni706en/ni706en.pdf
2. https://www.un.org/sustainabledevelopment/climate-change
3. Nordgren, A. (2022). Artificial intelligence and climate change: ethical issues. *Journal of Information Communication and Ethics in Society, 10*(1108). https://doi.org/10.1108/JICES-11-2021-0106
4. Janiesch, C., Zschech, P., & Heinrich, K. (2021). Machine learning and deep learning. *Journal of Electron Markets, 31*, 685–695. https://doi.org/10.1007/s12525-021-00475-2
5. Çabuk, S. (2019). Yolcuların Yeşil Havayollarına Ynelik Tutumları (Attitudes of passengers towards green airlines). *Journal of Yasar University, 14*, 237–250. https://www.researchgate.net/publication/335586651

6. Tiwari, T., Tiwari, T., & Tiwari, S. (2018). How artificial intelligence, machine learning and deep learning are radically different. *International Journals of Advanced Research in Computer Science and Software Engineering, 8*(2). https://doi.org/10.23956/ijarcsse.v8i2.569
7. Zhironkin, S., & Cehlár, M. (2022). Green economy and sustainable development: The outlook. *Journal of Energies, 15*(3), 1167. https://doi.org/10.3390/en15031167
8. https://www.tutorialspoint.com/machine_learning_with_python/machine_learning_with_python_tutorial.pdf
9. https://www.section.io/engineering-education/introduction-to-random-forest-in-machine-learning/
10. Sen, J., & Mehtab, S. (2022). Machine learning: Algorithms, models and applications. IntechOpen. https://doi.org/10.5772/intechopen.94615
11. https://stats.oecd.org/Index.aspx?DataSetCode=AIRTRANS_CO2
12. Kaewunruen, S., Sresakoolchai, J., Xiang, Y. (2021). Identification of weather influences on flight punctuality using machine learning approach. *Journal of Climate, 9*(8). https://www.mdpi.com/2225-1154/9/8/127

The Impact of Artificial Intelligence on Waste Management for Climate Change

Heba Alshater⊙, Yasmine S. Moemen, and Ibrahim El-Tantawy El-Sayed

1 Introduction

Expanded trash generation results from rapid urbanization, population growth, and global financial development. "To the World Bank (2018)" [1, 2], In developing countries, the amount of garbage generated yearly is predicted to rise significantly due to population increase and urbanization. According to a study, about a third of manufactured rigid trash was dangerously handled and organized in illegal rubbish dumps or unmonitored landfills. Furthermore, these squatter hones pose several environmental and health risks, like contaminating the groundwater, arriving at disintegration, increasing the cancer rate, decreasing childhood survival rates and causing birth anomalies [3]. That's why environmental deterioration necessitates the scientific community's focus on developing and evaluating new methods for managing it [4]. Toxic and harmful to the environment, solid matter generated from many sources

H. Alshater (✉)
Department of Forensic Medicine and Clinical Toxicology, University Hospital, Menoufia University, Shebin El-Kom, Egypt
e-mail: Heba.alshater@med.menofia.edu.eg
URL: http://www.egyptscience.net

Y. S. Moemen
Clinical Pathology Department, National Liver Institute, Menoufia University, Menoufia, Egypt
e-mail: yasmine_moemen@liver.menofia.edu.eg
URL: http://www.egyptscience.net

I. E.-T. El-Sayed
Chemistry Department, Faculty of Science, Menoufia University, Menoufia, Egypt
URL: http://www.egyptscience.net

H. Alshater · Y. S. Moemen · I. E.-T. El-Sayed
Scientific Research Group in Egypt (SRGE), Giza, Egypt

© The Author(s), under exclusive license to Springer Nature Switzerland AG 2023
A. E. Hassanien and A. Darwish (eds.), *The Power of Data: Driving Climate Change with Data Science and Artificial Intelligence Innovations*, Studies in Big Data 118,
https://doi.org/10.1007/978-3-031-22456-0_3

such as private and mechanical are usually referred to as "strong waste," and the environment in various categories takes after it, such as domestic, mechanical, commercial and dangerous, and biomedical [5]. People who care about the environment and researchers are putting together new studies to determine the negative effects of urban trash on the biological system. They've implemented effective waste management frameworks in various parts of the world to use wastes as assets via fabric recuperation in a feasible way [6–8]. The increased use of electronic devices over the last few decades has led to a growth in electronic waste due to the increasing inventive boom. Industrialization and increased health care and therapy services have led to the generation of hazardous and biological wastes [9–11]. Strong economic growth, particularly in China, has also spurred the evolution of cities where an increase in the need for energy and transportation has led to a major discussion about. Carbon dioxide (CO_2), methane (CH_4) and nitrous oxide (N_2O) are all greenhouse gases that contribute to the troposphere's radiative forcing and influence global climate change, as well as invading on essential functions of land–water intelligence and conduits. Strong synergies between environmental and climate-related measures that would be achieved [12, 13]. Climate change is expected to significantly impact agriculture in the Mediterranean basin, which currently contributes to greenhouse gas emissions (GHG).

SOC development is widely regarded as a means to mitigate and adapt to climate change by increasing soil organic carbon (SOC). Research then used rural action data and present climate circumstances to build a demonstration model at the territorial scale that displayed GHG outflows from EOM preparation or storage before soil application while also linking a widely used dynamic SOC turnover model [14]. Ecosystems, including cropland, grassland and woodland under different soil types and climate circumstances, have been tested using this model, which is based on criteria such as soil type, temperature, moisture content and plant cover. Research also shows that the present focus on deep learning unspecialized hardware has resulted in lower prices for "hardware for utilization cases that are not immediately commercially feasible," making it more expensive for varied research. Furthermore, research [15, 16]. As AI-enabled frameworks present new avenues for advancement in efforts to combat climate catastrophe, it is both feasible and appealing. Artificial intelligence (AI) may include, for example, assessing the influence of natural arrangement interventions, such as carbon taxation and carbon exchanging frameworks, on climate change impacts of mechanical generating, construction and cargo transportation. However, adequate management can only improve maintainable and impartial AI-based replies. This means that AI researchers must develop criteria and standards for reporting the environmental impact of AI research initiatives. For the climate catastrophe to be addressed, we must look beyond AI-based solutions and consider various approaches that place emphasize values associated with people, society and nature, as well as technology itself. However, the dual nature of AI creates moral questions about reducing AI's emissions of nursery gas and using AI for relief and adjustment. In addition to these climate-related ethical concerns, AI raises a slew of additional ethical concerns, including as questions of protection and the potential

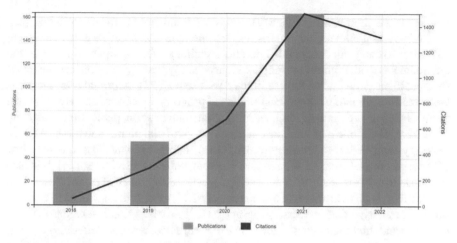

Fig. 1 Publications related to AI and energy management in the last five years according to the web of science database

danger of human tendencies [17]. Figure 1 depicts the most recent five years of AI and energy management papers according to the Web of Science database.

This section is organized as follows. Section 2 presents the application of AI in water management systems. Section 3 describes the role of Artificial Intelligence in estimating greenhouse gases. Section 4 presents the problems and challenges. Section 5 concludes this paper.

2 Waste Management and Artificial Intelligence Application

From the time the garbage is generated to its collection, transportation, and proper transfer to the time it is disposed of, waste management is one of the most pressing concerns faced by both developed and developing countries. Using it in collecting bins, transport vehicles and disposal methods is a viable option. use is also possible. Garbage management is also a major concern, as the increase in strong waste output is directly linked to the rise in sales and population growth, which includes an increasing number of customers. Lack of oversight of strong waste poses natural hazards and health risks, including contamination of groundwater and surface water, arrival weakening and other effects. The amount and quality of information available to environmental administrators are critical to effectively managing these concerns [18–20]. Studies show that poor waste management is primarily caused by a lack of effective planning and operational oversight [21, 22]. In recent years, real efforts have been made to overhaul the waste management business to make it more maintainable and productive using advanced technologies and innovative frameworks. Because of the numerous interconnected forms and the changeable statistical and financial elements that influence the overall frameworks, waste administration forms have complex operations and nonlinear characteristics. A difficult task in solid waste management (SWM) is to carry out acceptable operations without affecting the well-being of the environment or other natural resources. Consequently, researchers concentrate on AI methodologies well-suited for use in the SWM sector [23]. Artificial intelligence is concerned with creating computer systems and programs capable of emulating human abilities such as problem-solving, learning, conceptualization, perception, reasoning and awareness of one's immediate environment. Artificial neural networks (ANNs), expert systems (ESs), genetic algorithms (GAs), and fuzzy logic (FLs) are just a few of the many models that have been developed to address these issues [24]. It is also important to note that each artificial intelligence model or branch of it can be used for a specific task, such as artificial neural network (ANN) models that are used to classify, predict, or analyze data in urban geography. Contrariwise FL can safeguard human cognitive and thinking talents in addition to having a database and genetic algorithm) GA (receives the concept of common decision to obtain perfect results by selecting data best suited to deal with unexpected circumstances) [25]. Artificial intelligence-based risk management systems were also used to anticipate pollutant and particulate matter concentrations and to forecast atmospheric levels of carbon monoxide, ozone and nitrogen dioxide [26–28]. Water and wastewater treatment plant processes, waste generation patterns and garbage truck routes of garbage trucks, s were also examined [29, 30]. Since waste qualities can be accurately predicted, municipal solid waste may be collected and treated and disposed of methodically [4]. Various artificial intelligence (AI) methods were demonstrated by Yetilmezsoy et al. [24] to predict crash formation in terms of weight, content and rate. AI models used to forecast MSW generation rates in economic and socio-demographic variables have also been examined in other studies [31, 32]. Artificial intelligence-based optimization strategies in Solid Waste Management (SWM) have been debated by Melaré et al. [33] to predict waste generation, manage waste collection systems, monitor waste containers and select disposal sites [23]. Setting up a wide space for garbage collection and a designated place for dumping illegally discarded material is part of the waste management's efforts to minimize environmental human impact. ANNs were able to represent a wide range of SWM processes thanks to their robustness, fault tolerance and suitability for drawing complicated relationships between elements in multivariate frameworks. To make matters even better, the fewer parameters required by ANN frameworks' calibration handle means that these computations are much easier to perform. In addition, because of its low generalization cost and simple solution analysis, support vector machines (SVMs) have become widely connected within the SWM field [34]. There are many variables to consider while using SVMs, including the type of kernel and the tuning parameters. As a result, LR data analysis is used to model SWM, which depends on numerous characteristics. The results from LR are simple to interpret and need minimal processing resources. Another method for squandering classification and forecasting is decision tree analysis, which can also be utilized to identify waste behavior patterns and unlawful dump locations. Many SWM and AI models rely on genetic algorithms (GAs) or hybrid models to overcome the major constraints of current AI research. Many researchers are focusing on developing methods and processes to identify waste materials using automatic sorting order that reduces the need for manual garbage separation before applying AI approaches to make it easier to administrate, handle, transport and store solid waste. It is also clear that garbage transportation has a greater carbon footprint than the carbon footprint. However, artificial intelligence (AI) can help reduce carbon emissions by suggesting the most effective strategies [4]. SWM plans must include proper trash collection routes, which account for 70–85% of the overall cost of solid waste management. Patients and members of the general public should be kept from waste, and it must be ventilated and pest-free. The intelligent dumpster, equipped with AI programmers and IoT sensors, is a further step in garbage management. Because of this, rubbish may be measured by sensors on these dumpsters, transmitting this data to the main disposal system via intermediate computers for further processing and analysis. After receiving the alert, the garbage trucks and vans can then collect the waste from the overflowing bins. A modified Dijkstra algorithm in GIS was used to find the optimal approach options and the results were eventually implemented into GA by researchers in conjunction with GA and GIS. In addition, experts have shown that when garbage is collected through permitted channels, the contamination of hazardous substances from a variety of components is reduced. Stationary waste collections at municipal trash collection facilities, EEE merchants, mobile waste collections at curbside recycling stations and mobile terminals in high-traffic areas are all options for collecting waste. In addition, waste sorting robots have been put to use in landfills and the manual processes of waste sorting are being phased out in favor of intelligent automation. Researchers used nonlinear autoregressive neural networks and GIS route optimization to investigate the impact of garbage content and weight on optimal vehicle routes and emissions [4].

Mining operations, including exploration, extraction and transportation, need substantial vitality and greenhouse gas emissions. The type of vitality source used in any mining operation is largely determined by the mining technique and equipment used. Many of these energy and gas outflows can be decreased through enhanced monitoring. The mining sector is interested in conducting some studies on the fuel efficiency of portable machines in surface mining activities such as excavators, diggers and loaders because diesel is utilized as a source of power by these machines [35]. Traditional analytical approaches are commonly used to increase energy efficiency and unminimize emissions. However, typically fall short.

Further simple, rapid, low-cost solutions are needed to minimize gas emissions from surface mining. Machine learning and artificial intelligence are the best methods for accomplishing this (AI). Fuel consumption estimates for mine trucks are based on theoretical procedures and models produced by the truck manufacturer to achieve haul mining trucks, as is well known in this field of study. To better understand the relationship between haul truck payload, speed and total resistance (TR), the AI has focused on developing a model to estimate and reduce fuel consumption. This model was developed and tested in a surface coal mine in central Queensland, Australia, and the results were encouraging. For open-pit and open-cut mines, the improved model can estimate truck energy consumption using an artificial neural network (ANN) and Genetic Algorithm (GA) [36–42]. Many criteria can be utilized to determine a haul truck's fuel usage, including fleet management, mining planning, modern technology and the haul road [41, 43].

Employing global warming potential and represented in CO_2 equivalent (CO_2-e) or as a percentage of CO_2-e, diesel engine emissions can be computed [44–46]. The use of artificial neural systems (ANNs) is a widespread practice to simulate the impact of various variables on one crucial figure via a fitness function used to estimate fuel consumption by mining trucks, which considers some parameters that influence the consumption of fuel. Many engineering areas, such as materials, biological, and mechanical engineering, have used ANNs in their work [47–53]. A major advantage of ANNs is that they can simulate a wide range of nonlinear and linear relationships between variables by using the information given to the network to remember it as well as the representation of models employed by the brain to learn [53]. The advanced ANN algorithm can be used to estimate truck fuel consumption as a function of P, S, and TR according to these steps: normalizing the input parameters between -1 and $+1$, calculating the E and F parameters for each node, and finally calculating the E and F parameters for each node. These mathematical techniques imitate some of the well-known distinctive of the standard nerve arrangement and draw on analogies of adaptive accepted learning.

Before optimizing haul truck fuel consumption, the best approach to discover the measurable answer to problems and to shed light on specialist topics is first to estimate the fuel consumption of trucks. Then, the study area and the aim function must be evaluated. It's a mathematical function that links each answer point to an actual value in the answer region in the research area, which considers all possibilities. Additionally, demonstrating optimization methods based on Manufactured Insights rather than heuristic search has reduced the stiffness issue. Engineers use heuristic principles to solve all technical difficulties since they are based on the experience and perception of the framework in which the research takes place.

This class of models is known as Natural Optimization Methods, and it is widely accepted that they are among the best models with random characteristic behavior. As computer power increased in the 1980s, applying these models to capacity and process optimization became possible, which was previously impossible with conventional models. A few unique heuristic models were produced in the 1990s by the finished algorithms, such as reconstructed tempering, swarm calculations, subterranean insect colony optimization, and genetic calculation. It is also worth noting that the genetic algorithms (GAs), was proposed as an abstract of biological development, relying on principles from natural evolution, are a very modern optimization model and don't use any information derivate. Because of this, they stand a good chance of escaping the immediate danger and rely on the direct correlation between technological advances and natural wonders [54]. If enough gases formed, the population will focus on an optimal answer to the specified problem, and the operations are based on selection, reproduction and mutation. They have been used in various design, logical and financial difficulties because of their ability to optimize numerous functions. Numerous mathematical criteria for optimization issues, the handling of some goal functions and limits, the periodicity of progress administrators and the adaptability to hybridize with domain-dependent heuristics are the most advantageous aspects of this approach. In this approach, a significant number of tests and tests will be required to ensure an optimal collection of values for these components. The value of each property must match the reality of mine destinations and truck operation confinements to provide feasible arrangements before it can be used. All individuals must be examined through generations if they are in the same division.

Programming in Python was then used to finish the advanced models of ANN and GA. This was done by writing computer code that builds fitness functions based on the completed models. Mine truck fuel consumption is a function of payload, truck speed and total resistance in the first stage of the algorithm, which uses these input parameters. Aside from that, the new function is sent into the computer code's genetic algorithm phase as an input, and then the model will help to specify the code's specifics and will provide the best parameters. The solid waste generated during a project's construction, reconstruction, decoration and demolition phases significantly impacts the environment and socio-economic effectiveness of the entire project. As a result, numerous investigations have determined how this advanced development innovation can be applied to minimize the waste generated [55]. Demonstration of the potential for prefabrication to reduce construction waste at the project level through an active demonstration of the choice of prefabrication process during the plan organization. Models such as this one (SD) were first developed by Forreste [56] and have since been improved by various researchers through the use of model analysis, development, validation and evaluation [57]. Last but not least, Ding et al. [58] used the SD approach to evaluate the inherent advantages of CW decrease at both the planning and development stages. In this way, the SD technique is suited for motivating the complexity of the CW management system, including different stakeholders and components.

3 The Application of Artificial Intelligence to Estimate Greenhouse Gas (GHG)

Reducing greenhouse gas (GHG) emissions and air pollution has recently come to light as a potential advantage. Reduced emissions from Chinese megacities have piqued global interest because of China's position as the world's largest CO_2 emitter and a substantial contributor to the global air pollution budget [12]. Twenty-eight percent of SO_2 and around 10% of black carbon emissions from combustion make up the bulk of the air pollution budget [59, 60]. To meet China's objective of reducing carbon intensity by 40–45% by 2020, local governments are aggressively researching solutions and embracing strict energy-saving measures [61]. In developing municipal air quality programs incorporating climate change policy, scientific advice has been taken into account in various ways. Several studies focused on the co-benefits of specific policies, such as cleaner production projects in the coke industry in

Taiyuan City [62, 63] on regional features, but they failed to address major emission sources like transportation sectors. A complete technology-based technique to analyze the advantages of any policy has been developed in this manner. And the policy scenarios developed by employing the technique do so by studying the policy-level emission-reduction options. Compared to the current GAINS model, which relied on sector-level analysis to identify the benefits, this approach is significantly more effective. The appraisal of coordinates shows that picks have been successfully employed in analyzing co-benefits and related to driving the major transaction on pollution control understandings in Europe [64, 65]. Sulfur Protocol, Gothenburg multi-pollutant protocol, and new adjustments to national emission ceilings were also discussed in this meeting.

When it comes to large-scale (territorial or national) applications of the GAINS model, the city-scale outflow computation and arrangement evaluation require a few further breakthroughs to move forward with the characterization of adjacent situations. AIM/Local Model [66] and the GAINS-City model have a more complete illustration of source classification and inclusive pollutants, so they are a more realistic description of measures to guide policy decisions in Asia, for example, the linear programming MARKAL energy system optimization model for Shanghai [67]. GHG emissions and aquatic values have also been controlled using different models based on the Bayesian Belief Network idea for making decisions [68]. The mitigation potential can be affected by climate change, and other techniques have been used to assess and anticipate GHG emissions from rural frameworks beneath differentiating environments [69]. These techniques are also utilized in predicting how various farming practices affect GHG emissions by resolving interactions between farming practices and primary drivers (climate, soil typhoon). As a starting point [70], the suggested default outflow components, or Tier I, are exceptionally common and can lead to significant errors under certain conditions. Other countries have used advanced Tier II emission factors for agricultural systems based on country-specific measurements and thus provide more precise emission estimates. The Tier II outflow components may not apply to climate change estimates because they were designed under specific climatic conditions. As a result, models were created by scientists. In addition to DNDC and Daycent [71], models like Denitrification and Decomposition [72] and Decent [71], have been used to predict GHG emissions over space and time for a wide range of soil, farm management, and climatic conditions in various countries around the world [73–75]. For important climatic regions worldwide, Smith and Bertaglia [73] developed outflow variables for changes in agricultural management and estimated a worldwide moderation potential by 2030 of around 6000 Tg CO_2 eq. y-1. It also incorporates crop growth processes with soil biogeochemical processes on a daily time step. It mimics essential processes related to N and C cycles in plant–sol systems, including mineralization, ammonia volatilization, nitrification, denitrification, nitrogen uptake and leaching. This model has been a useful tool for evaluating the impact of management strategies on nitrogen and carbon fate in agroecosystems. Version 9.3 of the DNDC includes improvements [74] to the models' ability to simulate C and N fluxes in cold climate zones. The significant upgrade to this model also allows the prediction of the same soil–water properties irrespective of residue layer. So, the input datasets needed to build a national DNDC input database at the eco-district level for the entire country of Canada came from many sources and were in various formats and levels of geographic precision. In the absence of recycling, upstream greenhouse gas emissions occur mostly owing to the acquisition and handling of virgin crude materials for item manufacturing, necessitating greater use of fossil fuel than recycled materials. Solid waste can be reused as soon as it is generated to reduce greenhouse gas emissions (GHG) and conserve energy. In most circumstances, this is preferable to burning with or without recuperation of vitality to reduce GHG emissions. As a result, the resource conservation advantages of replacing virgin assets with reused resources tend to outweigh the energy generation offsets of waste-to-energy advancements. Reusing, on the other hand, conserves landfill capacity and extends the useful life of already-existing facilities [76]. Waste management programs in Brazil programs developed a program to reduce emissions of gases generated in areas of solid waste disposal, which are used in financial, social and natural preprograms proto mote economic advancement based on the criteria set by Brazil's Intermenstrual Commission on Worldwide Waste Management (Intermenstrual Commission) Changes in the climate. Recyclable material collection and distribution, as well as public awareness-raising efforts, are offered by a recycling cooperative with an average of 36 positions (Fundação) [77]. Estimate the synergies and trade-offs of CH_4 with NOx and CO, as well as technical implementation methods, from 1990 to 2030 using various sequences of transient atmospheric chemistry transport model computations [78]. Methane emissions are regulated to the utmost extent possible, and pollutants are kept at their CLE levels at the beginning of the scenario, according to this model. Ozone reductions will improve air quality and reduce the amount of radiation sent into the atmosphere. As a result, any efforts to reduce greenhouse gas emissions are nullified.

Most agricultural development operations put human health in jeopardy, especially when it comes to the overuse of fertilizerssers and unhygienic working conditions. Environmental degradation due to human activities such as excessive urbanization, agricultural methods, industrialization and population growth can be found around the world. In addition, water shortages have made it increasingly difficult to regulate pollutants and enhance water quality. The Government and experts have spent a lot of time studying water pollution. In light of the major pollution and global water scarcity, river water quality must be urgently protected [79].

Due to the expanding industrial activity, population growth, the depletion of consumable water and other human-caused and climate-related factors, it was determined that more than 1.1 billion people are unable to access safe drinking water [80]. Because of this, removing these pollutants from the wastewater is critical to protect both the natural environment and the future of humans. To meet the high standards for water quality, it's a challenging challenge for natural scientists to figure out how to remove these harmful contaminants and develop poisons from water frameworks. Normal settings, influent stun, and wastewater treatment technology all contribute to the vulnerability and variability of the wastewater treatment framework and wastewater treatment is a crucial stage in pollutant reduction and the improvement of water environment quality [81]. As a result of agricultural practices that involve the overuse of nitrate-containing fertilizers and animal waste, nitrate is one of the most common contaminants discovered in groundwater around the world [82]. Anaerobic digestion of municipal primary and waste-activated sludge produces ammonia that is then converted into the groundwater-contaminating biosolids that are formed as a result of this process [83, 84]. The biological processes that act as a barrier between pollution sources and natural aquatic ecosystems can be used to remove pollutants like heavy metals from wastewater [85, 86].

Many technical sectors, including wastewater treatment, water quality improvement, river water quality modeling and water resource management, use AI because of its capacity to tackle actual problems [87]. ANNs, which imitate biological neurons, are used in single and combination strategies in using artificial intelligence (AI) in wastewater treatment research. In wastewater treatment trials, artificial neural networks (ANNs) are used to remove water supply pollutants [88]. Each neuron receives input signals from the preceding neuron and analyses them before passing out the output, which acts as input to another set of neurons [89]. To forecast the nitrogen concentration of the treated wastewater, an ANN model with a prediction accuracy of 90% was employed [90]. Moreover, they could accurately forecast the decomposition of photoinduced polycyclic aromatic hydrocarbons in seawater [91, 92]. ANNs from the RBF and MLP families and the FNN and WNN families can also be used to build models and simulate the wastewater treatment process. Artificial neural networks (ANNs), feedback loops (FLs), genetic algorithms (GAs) and support vector machines (SVMs) are just a few examples of common AI techniques. The MT model can be used to solve a continuous class. In most cases, the experimental data set was divided into three or two parts: training, validation and testing. The training data set is one of three inputs when a model is being developed and tested in a prediction step. The other two are for validating the model and for testing its predictions. Many models are used to evaluate, forecast, and optimize COD removal in WWTP biochemical and physical processes. According to Moral and colleagues, the effluent COD of the Iskenderun Wastewater Treatment Plant was predicted using an artificial neural network model with an R2 of 0.632. [93]. Wan et al. [94] developed ANFIS, with a minimum MAPE of 1.0% and an R2 value of 0.982, as well as a GA ANN and non-dominant sorting GA-II for multi-objective optimization of the digestion process, to predict COD removal at a full-scale paper mill wastewater treatment plant using anoxic versus oxic processes. It was discovered that the GA-ANN model outperformed the standard ANN model in numerous metrics, including MSE, root means square normalization error (RRMS, average absolute percentage error, and correlation coefficient (r) [95]. Simulating nanocomposite absorbents to remove phosphate from water yielded an R2 value of 0.99 when an ANN and GA model were combined [96]. Methylene blue and malachite green removal from water using MLP-ANN and RBF-ANN approaches can be replicated and optimized utilizing the MLP-ANN model's better predictions than other methods [97, 98] which is based on an MLP-ANN and RBF-ANN, the total efficacy of phosphate (TP) removal is 86%, 79.9%, and 93%, respectively, with COD in place of SBR [99]. The results showed that the MLP-ANN model had a higher R2 and lower RMSE than the RBF-ANN model. MLR and ANN-GA models were used to study the adsorption of triamide on carbon nanotubes. It was shown that the ANN model predicted adsorption efficiency more accurately than the MLR model. In the presence of thiosemicarbazone advanced chitosan, RSM was found to be less accurate at predicting Pb^{2+} removal than MLP-ANN [100]. MLP-ANN and ANN, as well as ANN and ANN, were able to express the overall hydraulic resistance in a cross-flow microfiltration and membrane bioreactor for wastewater treatment [101, 102]. Nadiri et al. discovered that the SCFL model had a higher R2 value of 0.960 and a lower MAPE of 4% than individual FL when forecasting the BOD effluent from the Tabriz Wastewater Treatment Plant [103]. Nonlinear models such as FFNN, ANFIS, and SVM were then utilized to estimate the efficacy of lowering BODeff from wastewater treatment procedures using the conventional MLR. The combined ARIMA and ORELM models were used to forecast BOD5, COD, TDS, and TSS, and the effluent BOD model had an R2 of 0.99 to fulfill the improved performance requirements [104]. A multiobjective PSO algorithm, developed and implemented [105] was used to find the optimal nitrate set value in wastewater treatment, as was a revised adaptive kernel function model, which could describe the dynamic processes of water quality and energy consumption in wastewater treatment facilities [105]. Once an improved Q-learning algorithm was created, it was used to design a unique optimization approach for a biological phosphorus removal system that employed an optimal strategy for controlling both anaerobic and aerobic processes [106]. The DO levels in the effluent were managed using a DM strategy to maximize the ASP. If energy savings were prioritized over the air quality, the airflow might be lowered by 15%. [107]. Energy usage in an applied model afterward constructed through dynamic modeling (DM) was reduced due to the reduced oxygen production from the aeration process [108]. Thus, the MNLR and HMMs achieved an accuracy of 84% in estimating the total amount of inorganic nitrogen in the wastewater treatment process. According to the study by Suchetana et al. Algal [109]. remediation efficiency of As^{3+} and As^{5+} from wastewater utilizing the RBF-ANN model and its application in the emulsion liquid membrane was found to have R2 values of 0.9998 [110, 111]. With the help of artificial neural networks (ANNs), the pumping system's performance could be maintained while a 10% reduction in energy usage was accomplished [112]. Adopting a model-free RL [113] in wastewater treatment plants saves operating costs while maintaining acceptable water quality. The model-based operational costs have decreased by 6% compared to Benchmark Simulation. It has been created to predict As^{3+} adsorption on cerium hydrochloride by reducing the costs of adsorbents. It's been a great success [110] a hybrid NF control framework, consisting of ANN, GA, and FL, was developed to further improve the accuracy of the control framework and attain real-time control targets for dealing with wastewater treatment process instabilities. Could make better short-term and long-term predictions using a hybrid model of ANFIS and GFO than we could with ANFIS alone [114]. A Kohonen self-organizing feature map neural network was utilized to construct a model that projected the performance of the wastewater treatment facility due to unanticipated organic overload [115]. In addition, an ES was designed to check the current state of a pilot wastewater treatment plant, analyze its operating patterns, provide trustworthy data, good encouragement for administrators and determine the acidification status of anaerobic wastewater treatment plants with FL [116, 117]. Adaptive software sensors for wastewater treatment facilities were proposed using FL and neural network theory, and their NF model's simulation effect was superior to that of BP-ANNs [118, 119]. One of Gensym Corp's real-time ESs, G2, was stratified using an AI-based automatic control system. The results revealed that hybrid AI technology extended another approach for running complex wastewater treatment processing equipment efficiently while reducing energy consumption during [120]. As a result of the study's use of multiobjective optimization, the researchers could better manage the wastewater treatment process than they could have done using conventional methods [121]. New neural network structures were designed to improve SVI forecast accuracy in organic wastewater treatment by applying information-oriented computations and an improved LM computation. Using the Mard-RCP with Granger causal, a new fault detection framework was developed and deployed in Benchmark Simulation Model 1 and a full-scale WWTP to accurately identify sensor defects, sludge thickening, and influent shocks [122]. An enhanced PFA setup based on variable frequency mutations for activated sludge wastewater treatment might correctly identify and diagnose process faults [123]. AI was also used to control membrane fouling in water and wastewater treatment. The decline in flow was artificially simulated in cross-flow microfiltration via an FFNN model that provided an exact prognosis and maximized information and administrator involvement. This information was connected to assist administrators in making strides in administrating and controlling wastewater treatment plants, per the FFNN model [124]. Aside from the high costs and environmental risks of traditional water treatment methods, researchers decided to employ the Water Purification Monitoring methodology to minimize these drawbacks. For a drinking water purification procedure, they used a microcontroller-based monitoring system. While doing this procedure, it was necessary to make use of this monitoring device. It also has an LCD indicating circuit from which data is collected and sent to the person in charge of maintenance and repair. The detecting device can analyze data to evaluate whether the water filtration elements are in good working order. It can also send a message or play a sound to customers if the water filtering elements are in a harmful state. In case the purification system's water pump needs to be turned off, it has a power-switching device that can do so. As a result, the filtering components' ability to remove pollutants can be evaluated after some period of operation. The monitoring device will sound an alarm and instantly shut down the power supply or reduce the pump's speed to stop the water replenishment if the filtration elements are closed. Additionally, if the filter pieces aren't updated after a certain [125, 126].

There are numerous techniques and systems of monitoring, each with its own set of rules and regulations. Water quality monitoring can be divided into manual and real-time online sensor node locations. Many existing methods and methodologies appear to be inadequate to achieve the goal. Water bodies have been restored and cleared of the accumulation of nutrients and organic compounds that can lead to quality degradation by various conventional approaches that have demonstrated remarkable results. The manual method predominates in developing countries. In addition, the real-time sensor node approach is not widely used in industrialized countries. It is necessary to collect and compare data on the various characteristics to determine the extent to which the resource has been exploited or polluted. It's important to note that each parameter has its mean Samples of water that can be taken from various locations and depths by utilizing ntainers manually inserted into the water. Various tests are carried out on the samples once they have been sent to the laboratory. It is possible to monitor water quality without relying on existing monitoring methods like manual water sampling and sensor node approach using WSN by utilizing recent breakthroughs in sectors like data analysis, unmanned surface vessels (USVs), and the Internet of Things (IoT) [127]. As a result, an engineering method for tracking and monitoring several water quality factors (pH, dissolved oxygen, temperature, conductivity, etc.). In addition, USVs have the benefit of allowing researchers to examine water quality in locations otherwise inaccessible to humans, such as those located at great distances.

Another advantage of a USV-based technique is that it minimizes the risk of missing an area of a river. This is achieved through a standard embedded system of hardware and software operating together [128]. A wireless sensor network built on water purification station serves as the system's primary control and monitoring center [129]. They can use this gadget to check the water's quality and then communicate that information to maintenance staff members with the proper credentials to access the web. The purification unit must be maintained by defining two nodes of sensing elements: a transmitter and a re-transmitter, no matter how much a shift in pH or TDS is detected. This message is sent to maintenance personnel to indicate that one or both measured values are outside acceptable ranges. The purifying system needs to be serviced, as indicated by this notice. Once the Wi-Fi module has been connected to the internet, the collected data will be sent to a global database for analysis. Once this database has determined, it will show this information at the IP address previously set on the internet [129].

4 Challenges in Waste Management and Artificial Intelligence

A discussion of the primary obstacles and prospects of using AI in Latin American municipal garbage management is provided herein. Latin American countries can benefit from studying Machine Learning themes such as information ascription and database integration despite the obstacles to the quality of the information. Furthermore, it has been discovered that the area has enormous potential to benefit from internet-based instruments, as the prevalence of web access is enormous and shows a steady increase in the future. If you're keen to know how AI-based instruments are used in less-favored countries, you'll have to wait till the technological and human resources are more readily available, and the rest of the world may be as well [130]. There are, unfortunately, very few details available about garbage contamination, and even if there were, it would be difficult or expensive, even with the traditional garbage inspection process. There are numerous opportunities and advantages to implementing trash management, including lowering costs, increasing supply chain and fabric stream efficiency, reducing waste, or even eliminating it. Squander administration innovation, stage and data collecting are key barriers to overcome in the early stages of the development of squander management innovation [131]. Other issues include generation and separation, collection, transfer, transportation, treatment, disposal and recycling. It educates urban waste management, researchers and policymakers on health and environmental issues. It was noted that a wide range of stake holders from local governments to homeowners to special cleaning companies were involved. Insufficient awareness efforts, a lack of neighborhood committee help and a bad response to accusations are among the issues cited for generation and separation [132]. Despite this, the study found that there was a lack of infrastructure (C1), insufficient funding (C2), cybersecurity issues and a lack of trust in AI [133]. As a result, AI and IoT advancements can transform the difficulties and provide viable answers to the issues faced by society and administrative authorities.

5 Conclusion

The solid waste generated by human activities contributes to climate change while also being harmed. Commercial, mechanical, agricultural, biomedical, electronic, and toxic wastes are all generated. During various life cycle phases, each of these wastes emits greenhouse gases (GHGs). Carbon dioxide and methane are released into the atmosphere during the collection and cremation of wastes (vitality usage transportation and furnaces). The interconnectedness of climate change and disasters that the management of solid waste is adversely affected due to the shifting levels of disturbance brought on by climate change in one or more of its manifestations. Reducing garbage sent to landfills is a primary goal of the National Action Plan for Climate Change's resource recovery and recycling initiatives. As well as water purification methods to maximize the use of resources and desalination management. However, there are several disadvantages to these approaches, not the least of which is the high cost of time, effort and money. To this end, a large number of scientists are working to eliminate environmentally friendly wastes using artificial intelligence (AI) technology that can forecast emissions reductions in areas where solid waste and air pollutants are generated, as well as recovery of clean water, energy and a variety of other resources from the waste stream. Artificial intelligence (AI) has the potential to be a formidable tool in the fight against climate change. Consequently, AI frameworks and their ability to autonomously and remotely control devices have captured public interest. There is a great deal of potential for AI to be used for the good of humanity and the environment. Climate change, biodiversity conservation, ocean health, water management and contamination are some of the environmental issues that AI can help us address with new and innovative solutions.

References

1. World Bank. (2018). *What a waste 2.0: A global snapshot of solid waste management to 2050.* International Bank for Reconstruction and Development.
2. Abdallah, M. A., Talib, M. A., Feroz, S., Nasir, Q., Abdalla, H., & Mahfood, B. (2020). Artificial intelligence applications in solid waste management: A systematic research review. *Waste Management, 109*, 231–246.
3. Triassi, M., Alfano, R., Illario, M., Nardone, A., Caporale, O., & Montuori, P. (2015). Environmental pollution from illegal waste disposal and health effects: A review on the Triangle of Death. *International Journal of Environmental Research and Public Health, 12*, 1216–1236.
4. Sharma, P., & Vaid, U. (2021). Emerging role of artificial intelligence in waste management practices. *IOP Conference Series.: Earth Environmental Science, 889*, 012047.
5. Pant, H. (2022, June). Waste to energy Nepal, thesis for: Masters in urban and rural planning advisor: Paolo Vincenzo Genevose.
6. Kearney, V., Chan, J. W., Valdes, G., Solberg, T. D., & Yom, S. (2018). The application of artificial intelligence in the IMRT planning process for head and neck cancer. *Oral Oncology, 87*, 111–116.
7. Aristodemou, L., & Tietze, F. (2018). The state-of-the-art on intellectual property analytics.
8. Seyedzadeh, S., Rahimian, F. P., Glesk, I., & Roper, M. (2018). Machine learning for estimation of building energy consumption and performance: a review. *Visualization in Engineering, 6*.
9. van Gent, P., Melman, T., Farah, H., van Nes, N., & van Arem, B. (2018). Multi-level driver workload prediction using machine learning and off-the-shelf sensors. *Transportation Research Record, 2672*, 141–152.
10. Love-Koh, J., Peel, A., Rejon-Parrilla, J. C., Ennis, K., Lovett, R., Manca, A., Chalkidou, A., Wood, H., & Taylor, M. (2018). The future of precision medicine: Potential impacts for health technology assessment. *Pharmacoeconomics, 36*, 1439–51.
11. Zhang, X., Chen, X., Wang, J., Zhan, Z., & Li, J. (2018). Verifiable privacy-preserving single-layer perceptron training scheme in cloud computing. *Soft Computing, 22*, 7719–7732.
12. Liu, F., Klimont, Z., Zhang, Q., Cofala, J., Zhao, L., Huo, H., Nguyen, B., Schöpp, W., Sander, R., Zheng, B., Hong, C., He, K., Amann, M., & Heyes, C. (2013). Integrating mitigation of air pollutants and greenhouse gases in Chinese cities: development of GAINS-City model for Beijing. *Journal of Cleaner Production, 58*, 25–33.
13. Wilcock, R., Elliott, S., Hudson, N., Parkyn, S., & Quinn, J. (2008). Climate change mitigation for agriculture: water quality benefits and costs. *Water Science and Technology—WST, 58*(11).
14. Pardoa, G., del Pradoa, A., Martínez-Menab, M., Bustamantec, M. A., Rodríguez Martínd, J. A., Álvaro-Fuentese, J., & Moralc, R. (2017). Orchard and horticulture systems in Spanish Mediterranean coastal areas: Is there a real possibility to contribute to C sequestration. *Agriculture, Ecosystems and Environment, 238*, 153–167.
15. Taddeo, M., Tsamados, A., Cowls, J., & Floridi, L. (2021). Artificial intelligence and the climate emergency: Opportunities. *Challenges and Recommendations, 4*(6), 776–779.
16. Hooker, S. (2020). The hardware lottery. ArXiv200906489 Cs.
17. Nordgren, A. (2022). Artificial intelligence and climate change: ethical issues. *Journal of Information, Communication and Ethics in Society Emerald Publishing Limited*, 1477–996X.
18. Green, C. P., Engkvist, O., & Pairaudeau, G. (2018). The convergence of artificial intelligence and chemistry for improved drug discovery. *Future Medicinal Chemistry, 10*, 2573–2576.
19. Udias, A., Pastori, M., Dondeynaz, C., Carmona Moreno, C., Ali, A., Cattaneo, L., & Cano, J. (2018). A decision support tool to enhance agricultural growth in the Mékrou river basin (West Africa). *Computers and Electronics Agriculture, 154*, 467–481.
20. Giuffrida, M. V., Doerner, P., & Tsaftaris, S. A. (2018). Pheno-Deep counter: A unified and versatile deep learning architecture for leaf counting. *The Plant Journal, 96*, 880–890.
21. Hannan, M. A., Arebey, M., Begum, R. A., Mustafa, A., & Basri, H. (2013). An automated solid waste bin level detection system using Gabor wavelet filters and multilayer perception. *Resources, Conservation and Recycling, 72*, 33–42.

22. Malakahmad, A., & Khalil, N. D. (2011). Solid waste collection system in Ipoh city a review. In *International conference on business, engineering and industrial applications (ICBEIA)*, (pp. 174–179).
23. Vitorino, A., Melaré, D. S., Montenegro, S., Faceli, K., & Casadei, V. (2017). Technologies and decision support systems to aid solid-waste management: A systematic review. *Waste Management, 59*, 567–584.
24. Yetilmezsoy, K., Ozkaya, B., & Cakmakci, M. (2011). Artificial intelligence-based prediction models for environmental engineering. *Neural Network World, 3*, 193–218.
25. Kalogirou, S. A. (2003). Use of genetic algorithms for the optimal design of flat plate solar collectors. In *Proceedings of the ISES, solar world congress on solar energy for a sustainable future*, (pp. 14–19).
26. Roy, S. (2012). Prediction of particulate matter concentrations using artificial neural network. *Resource Environment, 2*, 30–36.
27. Shu, H., Lu, H., Fan, H., Chang, M., Shu, H., Lu, H., Fan, H., Chang, M., & Chen, J. (2006). Prediction for energy content of Taiwan municipal solid waste using multilayer perceptron neural networks. *Journal of the Air and Waste Management Association, 56*, 852–858.
28. Agirre-basurko, E., Ibarra-berastegi, G., & Madariaga, I. (2006). Regression and multi-layer perceptron-based models to forecast hourly O_3 and NO_2 levels in the Bilbao area. *Environmental Modelling and Software, 21*, 430–446.
29. Cakmakci, M. (2007). Adaptive neuro-fuzzy modelling of anaerobic digestion of primary sedimentation sludge. *Bioprocess and Biosystems Engineering, 30*, 349–357.
30. Chun, M., Kwak, K., & Ryu, J. (1999). Application of ANFIS for coagulant dosing process in a water purification plant. In *IEEE international fuzzy systems conference proceedings*, (pp. 1743–1748).
31. Goel, S., Ranjan, V. P., & Bardhan, B. (2017). Forecasting solid waste generation rates. In D. Sengupta, & S. Agrahari, (Eds.), *Modelling trends in solid and hazardous waste management*, (pp. 35–63).
32. Kolekar, K. A., Hazra, T., & Chakrabarty, S. N. (2016). A review on prediction of municipal solid waste generation models. *Procedia Environmental Sciences, 35*, 238–244.
33. Melaré, A. V. S., González, S. M., Faceli, K., & Casadei, V. (2017). Technologies and decision support systems to aid solid-waste management: A systematic review. *Waste Management, 59*, 567–584.
34. Harrington, P. (2012). *Machine learning in action*. Manning Publications Co.
35. Soofastaei, A. (2018). The application of artificial intelligence to reduce greenhouse gas emissions in the mining industry. In *Green technologies to improve the environment on earth Intech open.*
36. Beckman, R. (2012). *Haul trucks in Australian surface mines* (pp. 87–96). Mine Haulage.
37. De Francia, M., et al. (2015). Filling up the tank. *Australasian Mining Review., 2*(12), 56–57.
38. Alarie, S., & Gamache, M. (2002). Overview of solution strategies used in truck dispatching systems for open pit mines. *International Journal of Surface Mining, Reclamation and Environment., 16*(1), 59–76.
39. Bhat, V. (1996). A model for the optimal allocation of trucks for solid waste management. *Waste Management and Research, 14*(1), 87–96.
40. Burt, C. N., & Caccetta, L. (2007). Match factor for heterogeneous truck and loader fleets. *International Journal of Mining, Reclamation and Environment, 21*(4), 262–270.
41. Nel, S., Kizil, M. S., & Knights, P. (2011). Improving truck-shovel matching. In *35th APCOM symposium*, (pp. 381–391). University of Wollongong, NSW, Australasian Institute of Mining and Metallurgy (AusImm).
42. Caterpillar. (2013). Caterpillar performance handbook, (10th edn, Vol. 2). US Caterpillar Company.
43. Soofastaei, A., et al. (2014). Payload variance plays a critical role in the fuel consumption of mining haul trucks. *Australian Resources and Investment, 8*(4), 63–64.
44. ANGA. (2013). In: Department of industry, climate change, science, research and tertiary education, editor, (pp. 326–341). National Greenhouse Accounts Factors. Australian Government.

45. Kecojevic, V., & Komljenovic, D. (2010). Haul truck fuel consumption and CO_2 emission under various engine load conditions. *Mining Engineering, 62*(12), 44–48.
46. Kecojevic, V., & Komljenovic, D. (2011). Impact of Bulldozer's engine load factor on fuel consumption, CO_2 emission and cost. *American Journal of Environmental Sciences, 7*(2), 125–131.
47. Hammood, A. (2012). Development artificial neural network model to study the influence of oxidation process and zinc-electroplating on fatigue life of gray cast iron. *International Journal of Mechanical and Mechatronics Engineering, 12*(5), 128–136.
48. Xiang, L., Xiang, Y., & Wu, P. (2014). Prediction of the fatigue life of natural rubber composites by artificial neural network approaches. *Materials and Design, 57*(2), 180–185.
49. Sha, W., & Edwards, K. (2007). The use of artificial neural networks in materials science-based research. *Materials and Design, 28*(6), 1747–1752.
50. Talib, A., Abu Hasan, Y., & Abdul Rahman, N. (2009). Predicting biochemical oxygen demand as indicator of river pollution using artificial neural networks. In *18th World IMACS/MODSIM Congress*, (pp. 195–202).
51. Ekici, B., & Aksoy, T. (2009). Prediction of building energy consumption by using artificial neural networks. *Advances in Engineering Software, 40*(5), 356–362.
52. Beigmoradi, S., Hajabdollahi, H., & Ramezani, A. (2014). Multiobjective aero acoustic optimisation of rear end in a simplified car model by using hybrid robust parameter design, artificial neural networks and genetic algorithm methods. *Computers and Fluids, 90*, 123–132.
53. Rodriguez, J., et al. (2013). The use of artificial neural network (ANN) for modeling the useful life of the failure assessment in blades of steam turbines. *Engineering Failure Analysis, 35*, 562–575.
54. Lim, A. H., Lee, C.-S., & Raman, M. (2019). Hybrid genetic algorithm and association rules for mining workflow best practices. *Expert Systems with Applications, 39*(12), 10544–10551.
55. Yuan, R., Guo, F., Qian, Y., Cheng, B., Li, J., Tang, X., & Peng, X. (2022). A system dynamic model for simulating the potential of prefabrication on construction waste reduction. *Environmental Science and Pollution Research, 29*, 12589–12600.
56. Forreste, J. W. (1968). Theory series II industrial dynamics—after the first decade. *Management Science, 14*(7), 398–415.
57. Faezipour, M., & Ferreira, S. (2018). A system dynamics approach for sustainable water management in hospitals. *IEEE Systems Journal, 12*(2), 1278–1285.
58. Ding, Z. K., Zhu, M., Tam, V. W. Y., Yi, G., & Tran, C. N. N. (2018). A system dynamics-based environmental benefit assessment model of construction waste reduction management at the design and construction stages. *Journal of Cleaner Production, 176*, 676–692.
59. Smith, S. J., van Aardenne, J., Klimont, Z., Andres, R. J., Volke, A., & Delgado Arias, S. (2011). Anthropogenic sulfur dioxide emissions: 1850e2005. *Atmospheric Chemistry and Physics, 11*, 1101–1116.
60. Bond, T. C., Bhardwaj, E., Dong, R., Jogani, R., Jung, S., Roden, C., Streets, D. G., & Trautmann, N. M. (2007). Historical emissions of black and organic carbon aerosol from energy-related combustion, 1850–2000. *Global Biogeochemical Cycles, 21*, GB2018.
61. State Council Office Announcement, State Council Standing Committee Investigation and Decision on National Greenhouse Gas Emissions Reduction Control Target. (2009).
62. Aunan, K., Fang, J., Vennemo, H., Oye, K., & Seip, H. M. (2004). Co-benefits of climate policy e lessons learned from a study in Shanxi, China. *Energy Policy, 32*, 567–581.
63. Mestl, H. E. S., Aunan, K., Fang, J., Seip, H. M., Skjelvik, J. M., & Vennemo, H. (2005). Cleaner production as climate investment e integrated assessment in Taiyuan City, China. *Journal of Cleaner Production, 13*, 57–70.
64. Schöpp, W., Amann, M., Cofala, J., Heyes, C., & Klimont, Z. (1998). Integrated assessment of European air pollution emission control strategies. *Environmental Modelling and Software, 14*, 1–9.
65. Amann, M., Bertok, I., Borken, J., Chambers, A., Cofala, J., Dentener, F., Heyes, C., Hoglund, L., Klimont, Z., Purohit, P., Rafaj, P., Schöpp, W., Texeira, E., Toth, G., Wagner, F., & Winiwarter, W. (2008). *GAINS-Asia: A tool to combat air pollution and climate change simultaneously; methodology*. International Institute for Applied Systems Analysis (IIASA).

66. AIM Project Team, AIM/Local: A user's guide. AIM Interim Paper, IP-02-01 (2002).
67. Gielen, D., & Changhong, C. (2001). The CO_2 emission reduction benefits of Chinese energy policies and environmental policies: A case study for Shanghai, period 1995–2020. *Ecological Economics, 39*, 257–270.
68. Oliver, R. M. & Smith, J. Q. (Eds). (1990). *Influence diagrams, belief nets and decision analysis.* Wiley.
69. Smith, W. N., Grant, B. B., Desjardins, R. L., Worth, D., Li, C., Boles, S. H., & Huffman, E. C. (2010). A tool to link agricultural activity data with the DNDC model to estimate GHG, emission factors in Canada. *Agriculture, Ecosystems and Environment, 136*, 301–309.
70. IPCC. (2006). IPCC guidelines for national greenhouse gas inventories. In: S. Eggleston, L. Buendia, K. Miwa, T. Ngara, & K. Tanabe, (Eds.), *Agriculture, forestry and other land use*, (Vol. 4). IGES.
71. Del Grosso, S. J., Parton, W. J., Mosier, A. R., Hartman, M. D., Brenner, J., Ojima, D. S., & Schimel, D. S. (2001). Simulated interaction of carbon dynamics and nitrogen trace gas fluxes using the DAYCENT model. In M. Schaffer & L. M. Hansen (Eds.), *Modeling carbon and nitrogen dynamics for soil management* (pp. 303–332). CRC Press.
72. Li, C. (2000). Modeling trace gas emissions from agricultural ecosystems. *Nutrient Cycling in Agroecosystems, 58*, 259–276.
73. Smith, P., & Bertaglia, M. (2007). Greenhouse gas mitigation in agriculture. (Topic Editor) In: C. J. Cleveland, (Ed.), *Encyclopedia of earth.* Environmental Information Coalition. National Council for Science and the Environment.
74. Grant, B., Smith, W. N., Desjardins, R. L., Lemke, R., & Li, C. (2004). Estimated N_2O and CO_2 emissions as influenced by agricultural practices in Canada. *Climatic Change, 65*, 315–332.
75. Desjardins, R. L., Smith, W., Grant, B., Campbell, C., & Riznek, R. (2004). Management strategies to sequester carbon in agricultural soils and mitigation greenhouse gas emissions. *Climatic Change, 70*, 283–297.
76. King, M. F., & Gutberlet, J. (2013). Contribution of cooperative sector recycling to greenhouse gas emissions reduction: A case study of Ribeirão Pires, Brazil. *Waste Management, 33*, 2771–2780.
77. Fundação Nacional da Saúde–FUNASA. (2010). *Programas municipais de coleta seletiva de lixo como fator de sustentabilidade dos sistemas públicos de saneamento ambiental na região metropolitana de São Paulo.* Ministério da Saúde.
78. Dentener, F., Stevenson, D., Cofala, J., Mechler, R., Amann, M., Bergamaschi, P., Raes, F., & Derwent, R. (2005). The impact of air pollutant and methane emission controls on tropospheric ozone and radiative forcing: CTM calculations for the period 1990–2030. *Atmospheric Chemistry and Physics, 5*, 1731–1755.
79. Singh, M. R., & Gupta, A. (2016). Water pollution-sources, effects and control.
80. Adeleye, A. S., Conway, J. R., Garner, K., Huang, Y., Su, Y., & Keller, A. A. (2016). Engineered nanomaterials for water treatment and remediation: Costs, benefits, and applicability. *Chemical Engineering Journal, 286*, 640–662.
81. Zhaoa, L., Daia, T., Qiaoa, Z., Sun, P., Hao, J., & Yang, Y. (2020). Application of artificial intelligence to wastewater treatment: A bibliometric analysis and systematic review of technology, economy, management, and wastewater reuse. *Process Safety and Environmental Protection, 133*, 169–182.
82. Misiti, T. M., Hajaya, M. G., & Pavlostathis, S. G. (2011). Nitrate reduction in a simulated free-water surface wetland system. *Water Research*, 45, 5587–5598.
83. Surampalli, R. Y., Lai, K. C. K., Banerji, S. K., Smith, J., Tyagi, R. D., & Lohani, B. N. (2008). Long-term land application of biosolids e a case study. *Water Science and Technology, 57*(3), 345–352.
84. United States Environmental Protection Agency (US EPA), Biosolids Technology Fact Sheet e Land Application of Biosolids. (2000). Office of Water.
85. Bachand, P. A. M., & Horne, A. J. (2000). Denitrification in constructed free-water surface wetlands: I. Very high nitrate removal rates in a macrocosm study. *Ecological Engineering, 14*(1e2), 9–15.

86. Maltais-Landry, G., Maranger, R., Brisson, J., & Chazarenc, F. (2099). Nitrogen transformations and retention in planted and artificially aerated constructed wetlands. *Water Research, 43*(2), 535–545.

87. Xu, Y., Wang, Z., Jiang, Y., Yang, Y., & Wang, F. (2019). Small-world network analysis on fault propagation characteristics of water networks in eco-industrial parks. *Resources, Conservation and Recycling, 149*(343–351), 2019.

88. Fan, M., Hu, J., Cao, R., Ruan, W., & Wei, X. (2018). A review on experimental design for pollutants removal in water treatment with the aid of artificial intelligence. *Chemosphere, 200*, 330–343.

89. Chakraborty, T., Chakraborty, A. K., & Chattopadhyay, S. (2019). A novel distribution-free hybrid regression model for manufacturing process efficiency improvement. *Journal of Computational and Applied Mathematics, 362*, 130–142.

90. Chen, J. C., Chang, N. B., & Shieh, W. K. (2003). Assessing wastewater reclamation potential by neural network model. *Engineering Applications of Artificial Intelligence, 6*, 149–157.

91. Jing, L., Chen, B., Zhang, B., Zheng, J., & Liu, B. (2014). Naphthalene degradation in seawater by UV irradiation: The effects of fluence rate, salinity, temperature and initial concentration. *Marine Pollution Bulletin, 81*, 149–156.

92. Vakili, M., Mojiri, A., Kindaichi, T., Cagnetta, G., Yuan, J., Wang, B., & Giwa, A. S. (2019). Cross-linked chitosan/zeolite as a fixed-bed column for organic micropollutants removal from aqueous solution, optimization with RSM and artificial neural network. *Journal of Environmental Management, 250*, 109434.

93. Moral, H., Aksoy, A., & Gokcay, C. F. (2008). Modeling of the activated sludge process by using artificial neural networks with automated architecture screening. *Computers and Chemical Engineering, 32*, 2471–2478.

94. Wan, J., Huang, M., Ma, Y., Guo, W., Wang, Y., Zhang, H., Li, W., & Sun, X. (2011). Prediction of effluent quality of a paper mill wastewater treatment using an adaptive network-based fuzzy inference system. *Applied Soft Computing Journal, 11*, 3238–3246.

95. Huang, M., Han, W., Wan, J., Ma, Y., & Chen, X. (2016). Multiobjective optimization for design and operation of anaerobic digestion using GA-ANN and NSGA-II. *Journal of Chemical Technology and Biotechnology, 91*, 226–233.

96. Zhang, Y., & Pan, B. (2014). Modeling batch and column phosphate removal by hydrated ferric oxide-based nanocomposite using response surface methodology and artificial neural network. *Chemical Engineering Journal, 249*, 111–120.

97. Asfaram, A., Ghaedi, M., Hajati, S., Goudarzi, A., & Dil, E. A. (2017). Screening and optimization of highly effective ultrasound-assisted simultaneous adsorption of cationic dyes onto Mn-doped Fe_3O_4-nanoparticle-loaded activated carbon. *Ultrasonics Sonochemistry, 34*, 1–12.

98. Ranjbar-mohammadi, M., Rahimdokht, M., & Pajootan, E. (2019). Low-cost hydrogels based on gum Tragacanth and TiO_2 nanoparticles: Characterization and RBFNN modelling of methylene blue dye removal. *International Journal of Biological Macromolecules, 134*, 967–975.

99. Bagheri, M., Mirbagheri, S. A., Ehteshami, M., & Bagheri, Z. (2015). Modeling of a sequencing batch reactor treating municipal wastewater using multi-layer perceptron and radial basis function artificial neural networks. *Process Safety and Environment Protection, 93*, 111–123.

100. Peiman, S., Zaferani, G., Reza, M., Emami, S., Kiannejad, M., & Binaeian, E. (2019). Optimization of the removal Pb (II) and its Gibbs free energy by thiosemicarbazide modified chitosan using RSM and ANN modeling. *International Journal of Biological Macromolecules, 139*, 307–319.

101. Dornier, M., Decloux, M., Trystram, G., & Lebert, A. (1995). Dynamic modeling of cross-flow microfiltration using neural networks. *Journal of Membrane Science, 98*, 263–273.

102. Schmitt, F., & Do, K. U. (2017). Prediction of membrane fouling using artificial neural networks for wastewater treated by membrane bioreactor technologies: Bottlenecks and possibilities. *Environmental Science and Pollution Research, 24*, 22885–22913.

103. Nadiri, A. A., Shokri, S., Tsai, F. T. C., & Asghari Moghaddam, A. (2018). Prediction of effluent quality parameters of a wastewater treatment plant using a supervised committee fuzzy logic model. *Journal of Cleaner Production, 180*, 539–549.

104. Lotfi, K., Bonakdari, H., Ebtehaj, I., Mjalli, F. S., Zeynoddin, M., Delatolla, R., & Gharabaghi, B. (2019). Predicting wastewater treatment plant quality parameters using a novel hybrid linear-nonlinear methodology. *Journal of Environmental Management, 240*, 463–474.

105. Han, H. G., Zhang, L., Liu, H. X., & Qiao, J. F. (2018). Multiobjective design of fuzzy neural network controller for wastewater treatment process. *Applied Soft Computing Journal, 67*, 467–478.

106. Pang, J. W., Yang, S. S., He, L., Chen, Y. D., Cao, G. L., Zhao, L., Wang, X. Y., & Ren, N. Q. (2019). An influent responsive control strategy with machine learning: Q learning based optimization method for a biological phosphorus removal system. *Chemosphere, 234*, 893–901.

107. Kusiak, A., & Wei, X. (2012). Optimization of the activated sludge process. *Journal of Energy Engineering, 139*, 12–17.

108. Asadi, A., Verma, A., Yang, K., & Mejabi, B. (2017). Wastewater treatment aeration process optimization: A data mining approach. *Journal of Environmental Management, 203*, 630–639.

109. Suchetana, B., Rajagopalan, B., & Silverstein, J. A. (2019). Investigating regime shifts and the factors controlling total inorganic nitrogen concentrations in treated wastewater using non-homogeneous hidden Markov and multinomial logistic regression models. *Science of the Total Environment, 646*, 625–633.

110. Mandal, S., Mahapatra, S. S., Sahu, M. K., & Patel, R. K. (2015). Artificial neural network modelling of As(III) removal from water by novel hybrid material. *Process Safety and Environment Protection, 93*, 249–264.

111. Podder, M. S., Majumder, C. B. (2016). The use of artificial neural network for modelling of phycoremediation of toxic elements As(III) and As(V) from wastewater using Botryococcus braunii. *Spectrochimica Acta—Part A Molecular Biomolecular Spectroscopy, 155*, 130–145.

112. Zhang, Z., Kusiak, A., Zeng, Y., & Wei, X. (2016). Modeling and optimization of a wastewater pumping system with data-mining methods. *Applied Energy, 164*, 303–311.

113. Hernández-Del-Olmo, F., Llanes, F. H., & Gaudioso, E. (2012). An emergent approach for the control of wastewater treatment plants by means of reinforcement learning techniques. *Expert Systems with Applications, 39*, 2355–2360.

114. Dehghani, M., Seifi, A., & Riahi-Madvar, H. (2019). Novel forecasting models for immediate-short-term to long-term influent flow prediction by combining ANFIS and grey wolf optimization. *Journal of Hydrology, 576*, 698–725.

115. Çinar, Ö. (2005). New tool for evaluation of performance of wastewater treatment plant: Artificial neural network. *Process Biochemistry, 40*, 2980–2984.

116. Carrasco, E. F., Rodríguez, J., Pual, A., Roca, E., & Lema, J. M. (2002). Rule-based diagnosis and supervision of a pilot-scale wastewater treatment plant using fuzzy logic techniques. *Expert Systems with Applications, 22*(11–20), 2002.

117. Carrasco, E. F., Rodríguez, J., Punal, A., Roca, E., & Lema, J. M. (2004). Diagnosis of acidification states in an anaerobic wastewater treatment plant using a fuzzy based expert system. *Control Engineering Practice, 12*(59–64), 2004.

118. Huang, M. Z., Wan, J. Q., Ma, Y. W., Li, W. J., Sun, X. F., & Wan, Y. (2010). A fast predicting neural fuzzy model for online estimation of nutrient dynamics in an anoxic/oxic process. *Bioresource Technology, 10*, 1642–1651.

119. Mingzhi, H., Jinquan, W., Yongwen, M., Yan, W., Weijiang, L., & Xiaofei, S. (2009). Control rules of aeration in a submerged biofilm wastewater treatment process using fuzzy neural networks. *Expert Systems with Applications, 36*, 10428–10437.

120. Wen, C.-H., & Vassiliadis, C. A. (2002). Applying hybrid artificial intelligence techniques in wastewater treatment. *Engineering Applications of Artificial Intelligence, 11*, 685–705.

121. Dai, H., Chen, W., & Lu, X. (2016). The application of multiobjective optimization method for activated sludge process: A review. *Water Science and Technology, 76*, 223–235.

122. Cheng, H., Wu, J., Liu, Y., & Huang, D. (2019). A novel fault identification and root causality analysis of incipient faults with applications to wastewater treatment processes. *Chemometrics and Intelligent Laboratory Systems, 188*, 24–36.

123. Yu, P., Cao, J., Jegatheesan, V., & Shu, L. (2019). Activated sludge process faults diagnosis based on an improved particle filter algorithm. *Process Safety and Environment Protection, 127*, 66–72.

124. Mannina, G., Rebouças, T. F., Cosenza, A., Sànchez-Marrè, M., & Gibert, K. (2019). Decision support systems (DSS) for wastewater treatment plants—A review of the state of the art. *Bioresource Technology, 290*, 121814.

125. Khaleel, I. M. (2017). Ac to Ac frequency changer THD reduction based on selective harmonic elimination. *Journal of Electrical and Electronic Systems., 6*(1), 1–7.

126. Khesbak, M. S. M. A., & Khaleel, I. M. (2017). Novel multi tone- SPWM technique (MT-SPWM) using reference window and frequency optimization. In *2nd international conference on electrical, automation and mechanical engineering, advances in engineering research,* (Vol. 86, pp. 86–89).

127. Zhang, Q., Lan, Y., Chen, L., Yu, X., Zhang, L. (2021). Study of NB-IoT-based unmanned surface vehicle system for water quality monitoring of aquaculture ponds. In *Proceeding SPIE 11887, International conference on sensors and instruments,* (ICSI 2021).

128. Jan, F., Min-Allah, N., & Düştegör, D. (2021). IoT based smart water quality monitoring: recent techniques, trends and challenges for domestic applications. *Water, 13*, 1729.

129. Khalil, I. M., Abdulrazzak, H. N. (2019). Monitoring of water purification process based on IoT. *IOSR Journal of Electronics and Communication Engineering (IOSR-JECE), 14*(2), 56–62.

130. Bijos, J. C. B. F., Queiroz, L. M., Zanta, V. M., & Oliveira-Esquerre, K. P. (2021). Towards artificial intelligence in urban waste management: An early prospect for Latin America. *2021 International Conference on Resource Sustainability IOP Conference Series: Materials Science and Engineering,* (Vol. 1196, p. 012030).

131. Zaman, A. (2022). Waste management 4.0: An application of a machine learning model to identify and measure household waste contamination—A case study in Australia. *Sustainability, 14*, 3061.

132. Ebekozien, A., Aigbavboa, C., Emuchay, F. E., Aigbedion, M., Ogbaini, I. F., Awo-Osagie, A. I. (2021). Urban solid waste challenges and opportunities to promote sustainable developing cities through the fourth industrial revolution technologies. *International Journal of Building Pathology and Adaptation © Emerald Publishing Limited*, 2398–4708.

133. Wang, K., Zhao, Y., Gangadhari, R. K., & Li, Z. (2021). Analyzing the adoption challenges of the internet of things (IoT) and artificial intelligence (AI) for smart cities in China. *Sustainability, 13*.

A Machine Learning-Based Model for Predicting Temperature Under the Effects of Climate Change

Mahmoud Y. Shams⊙, Zahraa Tarek⊙, Ahmed M. Elshewey, Maha Hany, Ashraf Darwish⊙, and Aboul Ella Hassanien⊙

1 Introduction

The geology, biology, and ecosystems of world have undergone significant, perhaps irreversible changes as a result of climate change [1]. Numerous environmental dangers to human health have resulted from these developments, including the world-wide spread of infectious illnesses, strain on food production systems, loss of biodiversity, and ozone layer depletion [2]. Low-lying island nations in the Pacific will be most affected by rising sea levels and erosion, which will damage houses and

M. Y. Shams (✉)
Faculty of Artificial Intelligence, Kafrelsheikh University, Kafr El-Shaikh 33516, Egypt
e-mail: mahmoud.yasin@ai.kfs.edu.eg

Z. Tarek
Faculty of Computers and Information, Mansoura University, Mansoura 35516, Egypt
e-mail: zahraatarek@mans.edu.eg

A. M. Elshewey
Faculty of Computers and Information, Department of Computer Science, Suez University, Suez, Egypt
e-mail: elshewy86@gmail.com

M. Hany
Scientific City for Research and Technology Applications, Alexandria, Egypt
e-mail: mhany2005@yahoo.com

A. Darwish
Faculty of Science, Helwan University, Cairo, Egypt
e-mail: ashraf.darwish.eg@ieee.org

A. E. Hassanien
Faculty of Computers and Artificial Intelligence, Cairo University, Giza, Egypt
e-mail: aboitcairo@cu.edu.eg

M. Y. Shams · Z. Tarek · A. M. Elshewey · M. Hany · A. Darwish · A. E. Hassanien
Scientific Research Group in Egypt, Giza, Egypt

© The Author(s), under exclusive license to Springer Nature Switzerland AG 2023   61
A. E. Hassanien and A. Darwish (eds.), *The Power of Data: Driving Climate Change with Data Science and Artificial Intelligence Innovations*, Studies in Big Data 118,
https://doi.org/10.1007/978-3-031-22456-0_4

infrastructure and force people to evacuate. Rising ocean temperatures are linked to a rise and spread of illnesses in marine animals, and ingesting marine creatures puts people at risk of contracting these diseases directly [3, 4].

The process of monitoring climate change can be carried out through datasets that are collected to study its impact on the climate in general and provide recommendations for climate treatment. Artificial Intelligence (AI) can be a powerful tool in addressing some of humanity's greatest challenges. AI-driven systems help us to reduce the wasted energy amount in the home simply shutting off lighting and heating systems before we leave the home. Worldwide, these systems help combat drought by monitoring areas affected by desertification. Researchers are also looking about how data centers and AI computer systems themselves will affect the environment, such as finding a way to build systems and infrastructure that are more energy efficient. AI is just one tool in the complex process of analyzing the causes of climate change, but Its capacity to analyze massive volumes of data and find patterns enables us to comprehend the environment well. Machine learning enables AI systems to come up with solutions specific to those systems, rather than pre-programming them with a set of answers [5, 6].

Climate change is now a continuous event rather than something that will happen in the future. The reality of climate change has now been proved., and the changes create a greatest challenge for humanity today. So, climate change is the most burning issue now-a-days. It is evident that the global mean temperature has grown by 0.30–0.60 °C over last century. Due to a changing and disaster-prone climate, the global sea level has risen by 10–20 cm throughout the same time [7]. Egypt's climatic characteristics and weather pattern are undergoing some fundamental and significant changes. Temperatures are increasing, droughts and wildfires are becoming more common, rainfall patterns are changing, snow and glaciers are melting, and the globe sea level is increasing as a result of climate change. The emissions caused by human activity should be cut down on or completely stopped if we want to slow down climate change. Long-term changes in weather and temperature patterns are referred to as climate change. These changes might be caused by natural processes, such oscillations in the solar cycle. Since the 1800s, human activities—primarily the combustion of fossil fuels like oil, coal, and gas—have been the primary cause of climate change [8].

Forecasting is defined as the procedure of producing predictions based on early and current data. It is useful because it enables the creation of data-driven plans and the capacity to make educated judgments. Furthermore, it investigates how existing underlying currents imply prospective shifts in direction for enterprises, civilizations, or the entire planet. Prediction aims to provide future certainty. As a result, the primary goal of forecasting is to identify the whole range of possibilities rather than a small number of weak certainties. [9].

Air temperature prediction is the process of anticipating future temperature changes using a specific prediction model based on temperature time series data and other parameters. Temperature forecasting is used in weather forecasting, which can help to offer effective solutions to combat global warming. The ability to anticipate temperature variations is crucial for disaster warning, agricultural productivity, managing water resources, protecting the environment, and sustainable development. Recent years have seen an increase in the popularity of temperature forecasting on a worldwide scale [10].

The correlation between global warming and the increase in air temperature has lately caught experts' attention. The lives of people are eventually significantly impacted by climate change, which includes sea level rise, an increase in extreme occurrences, and global warming [11]. The atmosphere's state variable, air temperature, has an impact on both land surface and atmospheric processes. In order to safeguard people's lives and property, anticipating air temperature is a crucial component of weather prediction [12]. When the air temperature is outside of the acceptable range, people may experience possible health issues. Animals and plants may be harmed by abrupt temperature fluctuations. Due to the considerable impact that air temperature has on a number of industries, including agriculture, industry, and energy, an accurate prediction of air temperature is crucial. The precision of energy usage is improved by accurate air temperature forecasts [13]. Another important component in forecasting other meteorological variables is air temperature, including streamflow, solar radiation, and evapotranspiration. Finding a suitable method for air temperature forecasting is so essential and may help to lessen the effects of global warming and climate change. Furthermore, a strategy for human activities, economic growth, and energy policy, must take into account the precise forecast of air temperature [14–16]. The main contribution of this chapter is the studying the effect of climate change on the temperature by forecasting the global temperature using ML. We utilized standard dataset of climate change and perform preprocessing normalization for the selected features. Furthermore, we visualized the findings using both box and scatter plots. The evaluation results are performed using ML approaches; LR, RF KNN, DT, SVM, and CBR regressors.

The rest of the chapter is organized as follows. Section 2 briefly highlighted the current related work for predicting the global temperature. Section 3 presents the materials and method presented in this chapter. The proposed Cat Boost Regressor is presented in Sect. 4. The results and discussion are presented in Sect. 5. finally, the conclusion and perspectives are discussed in Sect. 6.

2 Related Work

Most research has focused on daily forecasting [17, 18], monthly [19] and annual mean temperatures [20, 21]. Hourly temperature forecasting has only received a small amount of research [22]. Traditional statistical methods including linear regression, cluster analysis, autoregressive integrated moving average (ARIMA), and grey prediction have been used to forecast air temperature up to this point. These models use statistical assessments of past data to determine the likelihood that a certain weather phenomenon will occur in the future. The mechanisms and variables driving variations in air temperature, on the other hand, are extremely intricate and nonlinear. Long time series of hourly or daily temperature cannot be successfully predicted using statistical methods because dynamic temperature changes are difficult to capture [9]. With the use of machine learning techniques like a support vector machine (SVM), the changing trend of air temperature has been anticipated [23], an artificial neural network (ANN) [24] and LSTM [25].

As an example, shallow learning approach, an SVM can estimate the maximum temperature for the following day across a time range of 2–10 days. An ANN, other shallow-learning technique, can accurately forecast changes in the daily average temperature trend [26]. A DBN and SAE, both deep learning techniques, are better at predicting temperature than a shallow neural network. A CNN (another deep learning approach) generates meteorological characteristics from convolution layers and sends them to a pooling layer, which selects and filters relevant information to limit the amount of input and prevent the CNN's gradient from disappearing [27]. As an extra deep learning approach, an RNN can anticipate time series of air temperature using neural units linked in a chain. A different deep-learning algorithm, LSTM, can anticipate short-term temperature correctly and effectively by gathering external input from hidden layers [28, 29].

To increase the accuracy of air temperature predictions, several deep-learning techniques have been included into models [30]. Convolutional recurrent neural network (CRNN) was created by combining a CNN with an RNN to identify the spatial and temporal correlations of the daily change in air temperature [31, 32].

Tran et al. [33] used standard LSTM, RNN, and multilayer ANN models. The highest air temperature in South Korea was predicted using hybrid models. The maximum air temperature for the next one to fifteen days was predicted using data from the previous seven days' worth of air temperature readings. The results demonstrated that the LSTM model outperformed the other models in terms of long-term air temperature predictions.

Tran and Lee make another attempt [34] utilized multilayer ANN models to forecast South Korea's maximum air temperature for one day. It was found that the ANN model generated the smallest error rate. For predicting air temperature, other research utilized deeper learning structures that were more complicated. As an example, Zhang et al. [31] used a convolutional recurrent neural network to predict the daily average air temperature for the next four days (CRNN). To train the CRNN, they used daily air temperature data collected across China between 1952 and 2018. Based on the historical air temperature data, the findings showed that their model could accurately forecast air temperature [26]. Cifuentes et al. [35] presented a deep learning approach to predict climate change temperature effect and they achieved minimum root mean square error 0.0017. As shown in Table 1, we summarize research that used neural network models to predict air temperature for a few minutes to many months in the future. Lin et al. [18] presents a temperature prediction based on multi-dimensional Empirical Decomposition Mode (EDM) ensemble and Radial Basis Function (RBF) with a minimum error rate in forecasting 7-day maximum temperature.

Table 1 The description of the current efforts to predict temperatures in presence of climate change

References	Input	Type of model	Output
Tran et al. [33]	Past daily maximum temperature (from 7 to 36)	Traditional ANN, RNN, LSTM	Daily maximum temperature for 1–15 days in advance
Tran and Lee [34]	Six previous daily maximum temperature	Traditional ANN	One day ahead maximum temperature
Zhang et al. [31]	Four past temperature data map series	Convolutional recurrent neural network (CRNN)	Four future temperature data map series
Cifuentes et al. [35]	Maximum, minimum, and average temperature, precipitation	Deep learning	Mean square error = 0.0017
Lin et al. [18]	Temperature, wind speed, humidity, and water depth are all factors to consider	Radial basis function neural network with multidimensional complementary ensemble empirical mode decomposition (MCEEMD) (RBFNN)	Forecast the daily maximum temperatures for the next seven days

3 Materials and Methods

The most significant environmental issue of the current era on a worldwide scale is climate change, which is brought on by global warming and is a consequence of rising greenhouse gas concentrations. Environmental issues brought on by climate change include shifting precipitation patterns, melting polar glaciers, and increasing sea levels [36]. Using a collection of climate change dataset, regression models were effective in predicting temperature in this chapter. This section provides a detailed discussion for the proposed model stages. The proposed model consists of data preprocessing for data normalization using Z normalization. After preparing and splitting the dataset, the regression models are applied to predict the temperature accurately. Different evaluation operators are used to evaluate the performance of the proposed prediction system. Results demonstrated that Cat boost regressor model achieved the best results with values of 0.003, 0.0036, 0.054, and 92.4% for MSE, MAE, RMSE, R^2, respectively. All steps of the proposed model are illustrated in Fig. 1.

Initially valid global dataset collected from the Climatic Research Unit at the University of East Anglia. Data was collected in a CSV file the data is normalized using Z normalization. Dataset is splitting to 70% for training, and 30% for testing. Multiple regression analyses were used interchangeably to predict temperature of the collected dataset related to climate change. The regression models e.g., Linear regression, Random regressor, K-nearest regressor, Decision tree regressor, Support

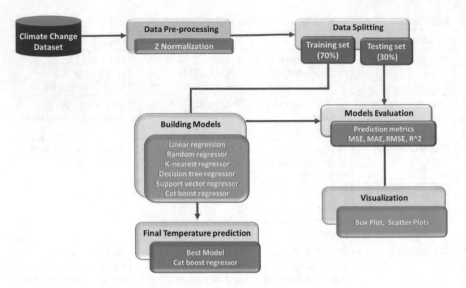

Fig. 1 The proposed temperature prediction model

vector regressor, have been trained with comparison to Cat boost regressor model for temperature prediction. Now, we'll discuss the regression models used in this chapter briefly.

3.1 Linear Regression

A well-known statistical technique, called linear regression, combines its input factors to predict the outcome variable in a straightforward and understandable manner [37]. One of linear regression's drawbacks is that it examines often a relationship between the mean of the output and input variables. Similar to how the mean does not adequately describe a single factor, linear regression does not provide a clear view of the relationships between variables [38]. Regression equation with numerous variables often look like as in Eq. (1):

$$Y = \beta + \beta_0 x_1 + \beta_1 x_2 + \beta_2 x_3 + \varepsilon. \tag{1}$$

where x_1, x_2, x_3 are the independent variables and Y is the predictor or target variable. $\beta_0, \beta_1, \beta_2$ are the coefficients and ε is the error term. Figure 2 presents the linear regression in machine learning [39].

Fig. 2 Linear regression in machine learning

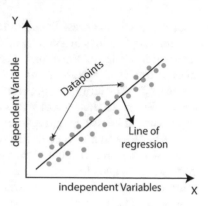

3.2 Decision Tree Regressor

One of the best regression techniques for handling high dimensional data is decision trees (DT). The decision trees are categories of non-parametric supervised learning techniques. The decision tree model learns "decision rules" derived from the properties of the current data to forecast the target variables. A piecewise constant approximation is similar to a decision tree regressor [40]. The decision tree divides recursively the feature space such that the samples with comparable target labels or values are grouped together.

Let Q_m with N_m samples represents the data at a node. Thus, the data is divided into $\mathbf{Q}_m^{left}(\theta)$ and $\mathbf{Q}_m^{right}(\theta)$ subsets for each split $\theta = (j, t_m)$ where j is a feature and t_m is a threshold [41], as shown in Eq. (2):

$$\mathbf{Q}_m^{left}(\theta) = \left\{ (x, y) | x_j \leq t_m \right\}$$
$$\mathbf{Q}_m^{right}(\theta) = Q_m / \mathbf{Q}_m^{left}(\theta) \tag{2}$$

The variable in Eq. (3) that minimizes the impurity (θ^*) is:

$$\theta^* = \operatorname{argmin}_\theta G(Q_m, \theta). \tag{3}$$

After the split's quality has been assessed, the parameters that reduce impurity denoted by θ^*.

There is recursively loop for subsets $\mathbf{Q}_m^{left}(\theta)$ and $\mathbf{Q}_m^{right}(\theta)$, until the maximum permitted depth is achieved with $N_m < \min_{samples}$, $N_m = 1$.

3.3 Random Forest Regressor

The utilization of an ensemble of trees is the most effective technique to improve DT's prediction accuracy. A collection of regression trees that were built at random is known as Random Forest. Random Forest Regressor (RFR) is more appropriate for real-time applications in a variety of domains, such as species distribution modeling, language modeling, bioinformatics, and ecosystem modeling [42]. Different decision trees are trained using subsets of the dataset, and the final results are determined using the concept of majority voting (averaging). Therefore, among its competitors, ANN and XGBoost, random forest was shown to be the optimal technique through the approaches of bootstrap aggregation and replacement [43].

In RFR, a forest is formed from a collection of N trees $\{R_1(X), R_2(X), ..., R_N(X)\}$, where X is a p-dimensional input vector with values $x_1, x_2, ..., x_p$. The ensemble generates N outputs for each tree $\hat{Y}_n = R_1(X), ..., \hat{Y}_N = R_N(X)$, where $\hat{Y}_n$ is the nth tree output and $n = 1, ..., N$. The average of each tree's unique forecasts makes up the final result. The bootstrap sample is used to build the tree. After training is complete, fully formed trees are utilized to forecast results for samples that have not been seen or are unidentified. The learning error provided by the Mean Square Error (MSE) is used to measure the predictive performance of Random Forests. Equation (4) shows that $\hat{Y}(X_i)$ is the anticipated output, Y_i is the observed output correlating to a certain input sample and n is the total number of out-of-bag samples [42].

$$MSE \approx MSE^{OOB} = n^{-1} \sum_{i=1}^{n} \left(\hat{Y}(Xi) - Y_i \right)^2. \tag{4}$$

3.4   K-Nearest Neighbor Regressor

The term "ML algorithms" refers to a type of algorithms that carry out computations implicitly through training rather than explicitly via programming. Most machine learning algorithms work by translating input characteristics to an output value (s). KNN are a class of ML algorithms that are reliable, easy to use, and cost little to run. The output value of a test point is specifically determined by KNN regressors as the interpolated value of the test point's closest neighbors. The number of neighbors to interpolate across, k, is used specified and the closest neighbors are determined by the k training points in the input feature space with the least Euclidean distance. Therefore, a weighted average of data points with comparable input properties may be used by KNN regressors to predict an output value [44]. Figure 3 illustrates the k-nearest neighbors regression [45].

3.5   Support Vector Regressor

SVR is a technique that may be used in regression situations when the support vector machine idea is being used. Due to its ability to address the overfitting issue, this

Fig. 3 k-nearest neighbors'
regression

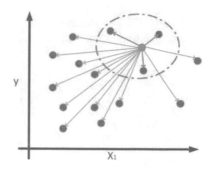

technique performs well in time series prediction and regression [46]. Equation (5)
is used to represent the function in the linear case.

$$f(x) = (w, x) + b \, with \, w \in x, b \in R \tag{5}$$

whereas w denotes the slope, x presents the feature space, and b indicates the inter-
cept. Equations (6) and (7) can be used in order to minimize the Euclidean value and
make the function as flat as feasible.

$$minimum \, \frac{1}{2} \left\| w^2 \right\| \tag{6}$$

depend on

$$\begin{cases} y_i - (w, x_i) - b \le \epsilon \\ (w, x_i) + b - y_i \le \epsilon \end{cases} \tag{7}$$

In contrast to SVM, which attempts to acquire a hyperplane by maximizing
margins and has a limit of 1 by following $y_i(w.x_i - b) \ge 1,$, SVR strives for regression to achieve the value with the least error or minimal margin (2ε), such that the
point is assumed to be inside the support hyperplane.

4 Cat Boost Regressor

Decision trees are used as the primary predictors in the majority of gradient boosting
implementations. Decision trees are useful for numerical characteristics, but in
reality, many datasets include categorical elements that are essential for prediction.
A cutting-edge gradient boosting technique is the Cat Boost Regressor (CAT). It
relates to more effective gradient boosting tree algorithmic framework implementation. A symmetrical decision tree algorithmic rule with categorical parameters,
minimal variables, and superior accuracy is the foundation of this framework [47].
The CatBoost method, like Gradient Boosting and XGBoost, builds several binary

decision trees every time it attempts to lower the error. The procedures for creating a tree in CatBoost are outlined in Algorithm 1 [48].

Algorithm 1: Constructing a tree in Cat Boost

input : $\{(x_p, y_p)\}_{p=1}^{n}$, M, L,α, $\{\sigma_s\}_{s=1}^{n}$,

Mod $\leftarrow$ Calc_Grad(L, M, y);

d$\leftarrow$ random(1, n);

if Mod = Plain **then**
  $\llcorner$ R $\leftarrow$ (grad$_r$(P) for P = 1..n);

if Mod = Ordered **then**
  $\llcorner$ R $\leftarrow$ (grad$_{r,\sigma r(P)-1}$(P) **for** P = 1..n);

V $\leftarrow$ vacuous tree;

foreach step from top to down **do**
  foreach candidate split c **do**
    S$_c$ $\leftarrow$ add split c to S;
    if Mod = Plain **then**
      $\llcorner$ Δ(p) $\leftarrow$ avgerage(grad$_r$(i) for i : leaf$_r$(i) = leaf$_r$(p)) for p = 1..n;
    if Mod = Ordered **then**
      $\llcorner$ Δ(p) $\leftarrow$ avgerage(grad$_{r,\sigma r(p)-1}$(i) for i : leaf$_r$(i) = leaf$_r$(p), σ_r(i) < σ_r(p)) for p = 1..n;
    $\llcorner$ loss(s$_c$) $\leftarrow$ cos(Δ, R)
  $\llcorner$ V $\leftarrow$ arg min$_{Tc}$ (loss(S$_c$))

if Mod = Plain **then**
  $\llcorner$ M_b(p) $\leftarrow$ M_b (p) $-$ α avgerage($grad_b$ (i) for i : $leaf_b$ (i) = $leaf_b$ (p)) for b = 1..s, p = 1..n;

if Mod = Ordered **then**
  $M_{b,r}$ (p) $\leftarrow$ $M_{b,r}$ (p) $-$ α avgerage($grad_{b,r}$ (i) for i : $leaf_b$ (i) = $leaf_b$ (p), σ_r (i) $\leq$ r)
  $\llcorner$ for b= 1..s, p= 1..n, r $\geq$ σ_b (p) $-$ 1;

return V, M

When compared to several machine learning models, the benefits of gradient boosting algorithms include their superior prediction accuracy. These methods offer a great deal of versatility, allowing for the optimization of the function fit utilizing a variety of hyperparameter tuning choices or other loss functions. They don't need to preprocess the data since they can operate directly with the input. The main disadvantage of gradient boosting techniques is their high computational cost. Overfitting may result from these systems' elimination of mistakes, and the parameters have a significant impact on how they behave [49].

The algorithm's accuracy and its generalizability are enhanced by CAT. Numerous areas, including media popularity prediction, weather forecasting, biomass, and evapotranspiration, have effectively utilized this algorithm [50]. It is for this reason that the model is applied here to predict the temperature in this chapter.

5 Evaluation Results and Discussion

5.1 Dataset Description

The dataset is available at https://www.kaggle.com/datasets/econdata/climate-change. The dataset consists of 308 instances and 11 features. The features names are Year, Month, Temp, CO_2, N_2O, CH_4, CFC.11, CFC.12, Aerosols, TSI, MEI. The climate data from May 1983 to December 2008. The statistical analysis and determination of the applied features are shown in Table 2. The statistical includes count, mean, standard deviation (std.), minimum (min), maximum, 25, 50, and 75% of the features. Figure 4 demonstrates the boxplot of the applied features of the predicted temperatures. While the scatter plot that describe the relationship between the temperature and the year is shown in Fig. 5.

Figure 6 shows the relation between the temperature and the aerosols. Furthermore, Fig. 7, discussed the relation between the temperature and the months for

Table 2 Statistical calculation for the features

	Year	Month	MEI	CO_2	CH_4	N_2O	CFC-11	CFC-12	TSI	Aerosols	Temp.
Count	308	308	308	308	308	308	308	308	308	308	308
Mean	1995	6.5	0.27	363	1749	312	251	497	1366	0.01	0.25
Std.	7.4	3.4	0.93	12	46	5.2	20	57	0.39	0.02	0.17
Min	1983	1	−1.6	340	1629	303	191	350	1365	0.001	−0.28
25%	1989	4	−0.39	353	1722	308	246	472	1365	0.002	0.12
50%	1996	7	0.23	361	1764	311	258	528	1365	0.005	0.24
75%	2002	10	0.83	373	1786	316	267	540	1366	0.012	0.40
Max	2008	12	3	388	1814	322	271	543	1367	0.14	0.73

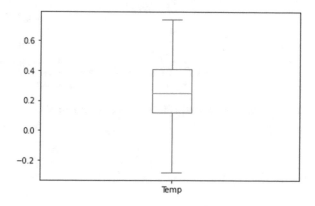

Fig. 4 Box plot for the applied temperature feature

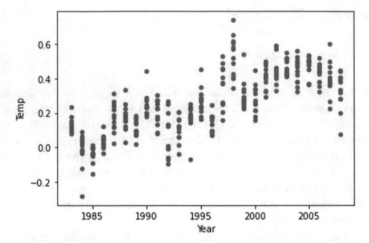

Fig. 5 Scatter plot between the temperature and years

one year. Figures 8, 9, 10, 11, 12, 13 and 14 describes the scatter plots relationship between the temperature and MEI, CO_2, CH_4, N_2O, CFC-11, CFC-12 and the TSI, respectively. The heatmap matrix that describes the relation and correlation between all the applied features (variables) of the dataset is shown in Fig. 15.

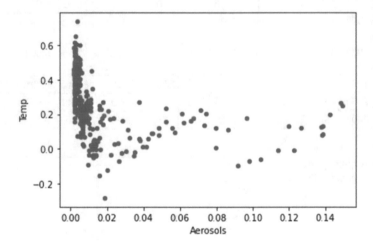

Fig. 6 Scatter plot between the temperature and aerosols

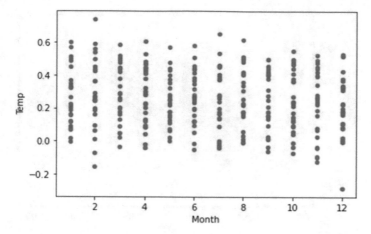

Fig. 7 Scatter plot between the temperature and month

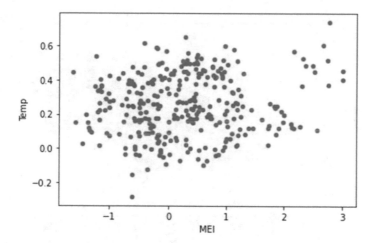

Fig. 8 Scatter plot between the temperature and MEI

5.2 Evaluation Results

The execution of the proposed regression ML (LR, RF KNN, DT, SVM, and CBR) models was assessed using the evaluation metrics, namely, Mean Square Error (MSE), Mean Absolute Error (MAE), Root Mean Square Error (RMSE), and determination coefficient (R^2). MSE is computed using Eq. (8):

$$MSE = \frac{1}{N} \sum_{i=1}^{N} \left(y_{actual_i} - y_{predicted_i} \right)^2 \tag{8}$$

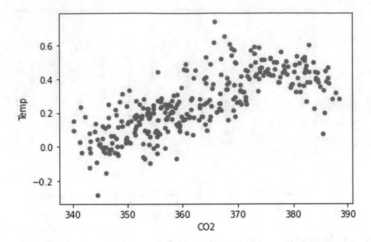

Fig. 9 Scatter plot between the temperature and CO_2

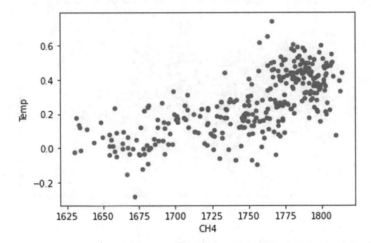

Fig. 10 Scatter plot between the temperature and CH_4

where N is the number of enrolled features and the y_{actual_i} is the actual values of the instance i. The predicted values of the instaces i is denoted by $y_{predicted_i}$. The RMSE, and the MAE are computed using Eqs. (9) and (10), respectively [51].

$$RMSE = \sqrt{\frac{1}{N} \sum_{i=1}^{N} \left(y_{actual_i} - y_{predicted_i} \right)^2} \tag{9}$$

$$MAE = \frac{1}{N} \sum_{i=1}^{N} \left| y_{actual_i} - y_{predicted_i} \right| \tag{10}$$

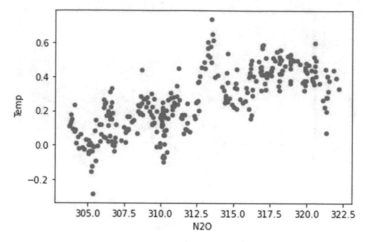

Fig. 11 Scatter plot between the temperature and N_2O

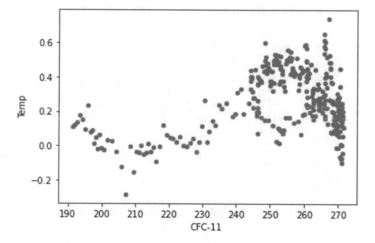

Fig. 12 Scatter plot between the temperature and CFC-11

The determination R^2 is computed using Eq. (11) as follows:

$$R^2 = 1 - \frac{\sum_{i=1}^{N} \left(y_{actual_i} - y_{predicted_i}\right)^2}{\sum_{i=1}^{N} \left(y_{actual_i} - \bar{y}\right)^2} \quad (11)$$

where $\bar{y}$ is the mean value. Table 3 demonstrates the results obtained based on the mentioned evaluation metrices for the ML (LR, RF KNN, DT, SVM, and CBR) regressor models. Figure 16 shows the error rates (MSE, MAE and RMSE) for the proposed models. We noticed that the minimum error rates achieved when using

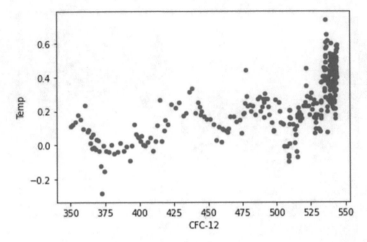

Fig. 13 Scatter plot between the temperature and CFC-12

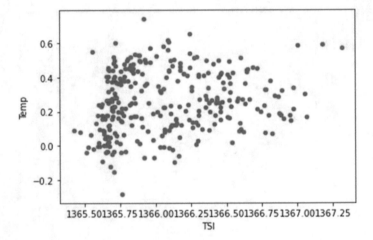

Fig. 14 Scatter plot between the temperature and TSI

Cat boot regressor. The results of determination coefficient (R^2) of the proposed ML models are shown in Fig. 17 indicated the superiority of the Cat boost regressor results (highlighted in bold) compared with other tradition ML regressor.

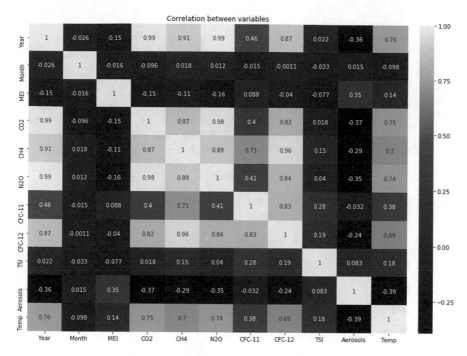

Fig. 15 The heatmap matrix for the features

Table 3 The results of the proposed ML models based on MSE, MAE, RMSE and R^2

Models	MSE	MAE	RMSE	R^2 (%)
Linear regression	0.009	0.0092	0.094	85.3
Random regressor	0.007	0.0075	0.787	88.9
K-nearest regressor	0.005	0.0058	0.070	90.6
Decision tree regressor	0.007	0.0071	0.787	88.4
Support vector regressor	0.004	0.0043	0.063	91.1
Cat boost regressor	**0.003**	**0.0036**	**0.054**	**92.4**

6 Conclusion and Perspectives

Climate change has a great effect on the surrounding environment. Currently the universe vulnerable great effect results from climate change. Especially the temperature which consider one of the important factors of climate change. This chapter present a prediction model based on ML regressors such as LR, RF KNN, DT, SVM, and CBR. Furthermore, we analysis the data using statistical analysis as well ae understanding the correlation between variables and the scatter plot relationship between the temperature and the other remaining features. These features are preprocessed and normalized using z-score normalization. Afterwards, it divided into 70% training

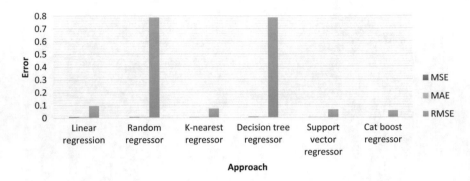

Fig. 16 The error values MSE, MAE and RMSE of the proposed ML models

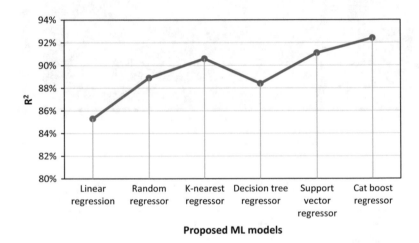

Fig. 17 The R^2 results compared with the proposed ML regressors

and the reminder 30% for testing. We used the prediction MSE, MAE, RMSE, and R^2 metrices. The results prove the ability of the proposed CBR model to predict the temperature with minimum error rates and high R^2 value compared with other ML models. In the future, we plan to utilize another climate change factor and recommend how can we tackle the change occurs. Moreover, we intended to study the effect of temperature on the agriculture fields especially the quality control of crops in the presence of climate change.

References

1. Steffen, W., Persson, Å., Deutsch, L., Zalasiewicz, J., & Williams, M., et al.: The Anthropocene: From global change to planetary stewardship. *AMBIO, 40*(7), 739 (2011). https://doi.org/10.1007/s13280-011-0185-x

2. Brierley, A. S., & Kingsford, M. J. (2009). Impacts of climate change on marine organisms and ecosystems. *Current Biology, 19*(14), R602–R614. https://doi.org/10.1016/j.cub.2009.05.046

3. Abraham, J. P., et al. (2013). A review of global ocean temperature observations: Implications for ocean heat content estimates and climate change. *Reviews of Geophysics, 51*(3), 450–483. https://doi.org/10.1002/rog.20022

4. Cheng, L., et al. (2021). Upper ocean temperatures hit record high in 2020. *Advances in Atmospheric Sciences, 38*(4), 523–530. https://doi.org/10.1007/s00376-021-0447-x

5. Fathi, S., Srinivasan, R. S., Kibert, C. J., Steiner, R. L., & Demirezen, E. (2020) AI-based campus energy use prediction for assessing the effects of climate change. *Sustainability, 12*(8), 8. https://doi.org/10.3390/su12083223

6. Chakraborty, D., Alam, A., Chaudhuri, S., Başağaoğlu, H., Sulbaran, T., & Langar, S. (2021). Scenario-based prediction of climate change impacts on building cooling energy consumption with explainable artificial intelligence. *Applied Energy, 291*, 116807. https://doi.org/10.1016/j.apenergy.2021.116807

7. Rahman, M. M., Opu, R. K., Sultana, A., & Riyad, M. R. A. (2020). *Climate change scenarios and analysis of temperature and rainfall intensity in Faridpur district*, Bangladesh

8. Hamed, M. M., Nashwan, M. S., & Shahid, S. (2022). Climatic zonation of Egypt based on high-resolution dataset using image clustering technique. *Progress in Earth and Planetary Science, 9*(1), 35. https://doi.org/10.1186/s40645-022-00494-3

9. Hou, J., Wang, Y., Zhou, J., & Tian, Q. (2022). Prediction of hourly air temperature based on CNN–LSTM. *Geomatics, Natural Hazards and Risk, 13*(1), 1962–1986. https://doi.org/10.1080/19475705.2022.2102942

10. Choi, B., Bergés, M., Bou-Zeid, E., & Pozzi, M. (2021). Short-term probabilistic forecasting of meso-scale near-surface urban temperature fields. *Environmental Modelling & Software, 145*, 105189. https://doi.org/10.1016/j.envsoft.2021.105189

11. Tajfar, E., Bateni, S. M., Lakshmi, V., & Ek, M. (2020). Estimation of surface heat fluxes via variational assimilation of land surface temperature, air temperature and specific humidity into a coupled land surface-atmospheric boundary layer model. *Journal of Hydrology, 583*, 124577. https://doi.org/10.1016/j.jhydrol.2020.124577

12. Tajfar, E., Bateni, S. M., Margulis, S. A., Gentine, P., & Auligne, T. (2020). Estimation of turbulent heat fluxes via assimilation of air temperature and specific humidity into an atmospheric boundary layer model. *Journal of Hydrometeorology, 21*(2), 205–225. https://doi.org/10.1175/JHM-D-19-0104.1

13. Valipour, M., Bateni, S. M., Gholami Sefidkouhi, M. A., Raeini-Sarjaz, M., & Singh, V. P. (2020). Complexity of forces driving trend of reference evapotranspiration and signals of climate change. *Atmosphere, 11*(10), 10. https://doi.org/10.3390/atmos11101081

14. Schulte, P. A., et al. (2016). Advancing the framework for considering the effects of climate change on worker safety and health. *Journal of Occupational and Environmental Hygiene, 13*(11), 847–865. https://doi.org/10.1080/15459624.2016.1179388

15. Marzo, A., et al. (2017). Daily global solar radiation estimation in desert areas using daily extreme temperatures and extraterrestrial radiation. *Renewable Energy, 113*, 303–311. https://doi.org/10.1016/j.renene.2017.01.061

16. Jovic, S., Nedeljkovic, B., Golubovic, Z., & Kostic, N. (2018). Evolutionary algorithm for reference evapotranspiration analysis. *Computers and Electronics in Agriculture, 150*, 1–4. https://doi.org/10.1016/j.compag.2018.04.003

17. Asha, J., Kumar, S. S., & Rishidas, S. (2021). Forecasting performance comparison of daily maximum temperature using ARMA based methods. *Journal of Physics Conference Series, 1921*(1), 012041. https://doi.org/10.1088/1742-6596/1921/1/012041

18. Lin, M.-L., Tsai, C. W., & Chen, C.-K. (2021). Daily maximum temperature forecasting in changing climate using a hybrid of multi-dimensional complementary ensemble empirical mode decomposition and radial basis function neural network. *Journal of Hydrology: Regional Studies, 38*, 100923. https://doi.org/10.1016/j.ejrh.2021.100923

19. Narasimha Murthy, K. V., Saravana, R., Kishore Kumar, G., & Vijaya Kumar, K. (2021). Modelling and forecasting for monthly surface air temperature patterns in India, 1951–2016: Structural time series approach. *Journal of Earth System Science, 130*(1), 21. https://doi.org/10.1007/s12040-020-01521-x

20. Liu, Z., et al. (2019). Balancing prediction accuracy and generalization ability: A hybrid framework for modelling the annual dynamics of satellite-derived land surface temperatures. *ISPRS Journal of Photogrammetry and Remote Sensing, 151*, 189–206. https://doi.org/10.1016/j.isprsjprs.2019.03.013

21. Johnson, Z. C., et al. (2020). Paired air-water annual temperature patterns reveal hydrogeological controls on stream thermal regimes at watershed to continental scales. *Journal of Hydrology, 587*, 124929. https://doi.org/10.1016/j.jhydrol.2020.124929

22. Carrión, D., et al. (2021). A 1-km hourly air-temperature model for 13 northeastern U.S. states using remotely sensed and ground-based measurements. *Environmental Research, 200*, 111477. https://doi.org/10.1016/j.envres.2021.111477

23. Gos, M., Krzyszczak, J., Baranowski, P., Murat, M., & Malinowska, I. (2020). Combined TBATS and SVM model of minimum and maximum air temperatures applied to wheat yield prediction at different locations in Europe. *Agricultural and Forest Meteorology, 281*, 107827. https://doi.org/10.1016/j.agrformet.2019.107827

24. Astsatryan, H., et al. (2021). Air temperature forecasting using artificial neural network for Ararat valley. *Earth Science Informatics, 14*(2), 711–722. https://doi.org/10.1007/s12145-021-00583-9

25. Bai, P., Liu, X., & Xie, J. (2021). Simulating runoff under changing climatic conditions: A comparison of the long short-term memory network with two conceptual hydrologic models. *Journal of Hydrology, 592*, 125779. https://doi.org/10.1016/j.jhydrol.2020.125779

26. Tran, T. T. K., Bateni, S. M., Ki, S. J., & Vosoughifar, H. (2021). A review of neural networks for air temperature forecasting. *Water, 13*(9), 9. https://doi.org/10.3390/w13091294

27. Bayatvarkeshi, M., et al. (2021). Modeling soil temperature using air temperature features in diverse climatic conditions with complementary machine learning models. *Computers and Electronics in Agriculture, 185*, 106158. https://doi.org/10.1016/j.compag.2021.106158

28. Mtibaa, F., Nguyen, K.-K., Azam, M., Papachristou, A., Venne, J.-S., & Cheriet, M. (2020). LSTM-based indoor air temperature prediction framework for HVAC systems in smart buildings. *Neural Computing and Applications, 32*(23), 17569–17585. https://doi.org/10.1007/s00521-020-04926-3

29. Sekertekin, A., Bilgili, M., Arslan, N., Yildirim, A., Celebi, K., & Ozbek, A. (2021). Short-term air temperature prediction by adaptive neuro-fuzzy inference system (ANFIS) and long short-term memory (LSTM) network. *Meteorology and Atmospheric Physics, 133*(3), 943–959. https://doi.org/10.1007/s00703-021-00791-4

30. Mohammadi, B., Mehdizadeh, S., Ahmadi, F., Lien, N. T. T., Linh, N. T. T., & Pham, Q. B. (2021). Developing hybrid time series and artificial intelligence models for estimating air temperatures. *Stochastic Environmental Research Risk Assessment, 35*(6), 1189–1204. https://doi.org/10.1007/s00477-020-01898-7

31. Zhang, Z., & Dong, Y. (2020). Temperature forecasting via convolutional recurrent neural networks based on time-series data. *Complexity, 2020*, e3536572. https://doi.org/10.1155/2020/3536572

32. Tabrizi, S. E., et al. (2021). Hourly road pavement surface temperature forecasting using deep learning models. *Journal of Hydrology, 603*, 126877. https://doi.org/10.1016/j.jhydrol.2021.126877

33. Thi Kieu Tran, T., Lee, T., Shin, J.-Y., Kim, J.-S., & Kamruzzaman, M. (2020). Deep learning-based maximum temperature forecasting assisted with meta-learning for hyperparameter optimization. *Atmosphere, 11*(5), 5. https://doi.org/10.3390/atmos11050487

34. Tran, T. T. K., Lee, T., & Kim, J.-S. (2020). Increasing neurons or deepening layers in forecasting maximum temperature time series? *Atmosphere, 11*(10), 10. https://doi.org/10.3390/atmos1110 1072

35. Cifuentes, J., Marulanda, G., Bello, A., & Reneses, J. (2020). Air temperature forecasting using machine learning techniques: A review. *Energies, 13*(16), 16. https://doi.org/10.3390/en1316 4215

36. Guo, L.-N., et al. (2021). Prediction of the effects of climate change on hydroelectric generation, electricity demand, and emissions of greenhouse gases under climatic scenarios and optimized ANN model. *Energy Reports, 7*, 5431–5445.

37. Liang, D., et al. (2022). Examining the utility of nonlinear machine learning approaches versus linear regression for predicting body image outcomes: The US body project I. *Body Image, 41*, 32–45.

38. Rath, S., Tripathy, A., & Tripathy, A. R. (2020). Prediction of new active cases of coronavirus disease (COVID-19) pandemic using multiple linear regression model. *Diabetes & Metabolic Syndrome: Clinical Research & Reviews, 14*(5), 1467–1474.

39. Javatpoint. (2021). Linear regression in machine learning. In *Javatpoint*. https://www.javatp oint.com/linear-regression-in-machine-learning. Accessed September 27, 2022.

40. An, Y., Wang, X., Qu, Z., Liao, T., & Nan, Z. (2018). Fiber Bragg grating temperature calibration based on BP neural network. *Optik, 172*, 753–759.

41. Dhanalakshmi, S., et al. (2022). Fiber Bragg grating sensor-based temperature monitoring of solar photovoltaic panels using machine learning algorithms. *Optical Fiber Technology, 69*, 102831.

42. Adusumilli, S., Bhatt, D., Wang, H., Devabhaktuni, V., & Bhattacharya, P. (2015). A novel hybrid approach utilizing principal component regression and random forest regression to bridge the period of GPS outages. *Neurocomputing, 166*, 185–192.

43. Mussumeci, E., & Coelho, F. C. (2020). Large-scale multivariate forecasting models for Dengue-LSTM versus random forest regression. *Spatial and Spatio-Temporal Epidemiology, 35*, 100372.

44. Durbin, M., Wonders, M. A., Flaska, M., & Lintereur, A. T. (2021). K-nearest neighbors regression for the discrimination of gamma rays and neutrons in organic scintillators. *Nuclear Instruments and Methods in Physics Research Section A: Accelerators, Spectrometers, Detectors and Associated Equipment, 987*, 164826.

45. Qaddoura, R., & Younes, M. B. (2022). Temporal prediction of traffic characteristics on real road scenarios in Amman. *Journal of Ambient Intelligence and Humanized Computing*, 1–16.

46. Kurniawan, R., Setiawan, I. N., Caraka, R. E., & Nasution, B. I. (2022). Using Harris hawk optimization towards support vector regression to ozone prediction. *Stochastic Environmental Research and Risk Assessment, 36*(2), 429–449.

47. Khan, P. W., Byun, Y.-C., Lee, S.-J., Kang, D.-H., Kang, J.-Y., & Park, H.-S. (2020). Machine learning-based approach to predict energy consumption of renewable and nonrenewable power sources. *Energies, 13*(18), 4870.

48. Prokhorenkova, L., Gusev, G., Vorobev, A., Dorogush, A. V., & Gulin, A. (2018). CatBoost: Unbiased boosting with categorical features. *Advances in Neural Information Processing Systems, 31*.

49. Nguyen, V.-H., et al. (2021). Applying Bayesian optimization for machine learning models in predicting the surface roughness in single-point diamond turning polycarbonate. *Mathematical Problems in Engineering, 2021*.

50. Zhang, Y., Ma, J., Liang, S., Li, X., & Li, M. (2020). An evaluation of eight machine learning regression algorithms for forest aboveground biomass estimation from multiple satellite data products. *Remote Sensing, 12*(24), 4015.

51. Shams, M. Y., Tolba, A. S., & Sarhan, S. H. (2017). A vision system for multi-view face recognition. *International Journal of Circuits, Systems, and Signal Processing, 10*(1), 455–461. arXiv:1706.00510

Emerging Technologies in Industry and Energy Sector

Prediction of CO_2 Emission in Cars Using Machine Learning Algorithms

Gehad Ismail Sayed and Aboul Ella Hassanien

Abstract Greenhouse gases such as carbon dioxide emissions or shortly CO_2 are considered one of the essential causes of climate change. Recently, it has been considered one of the most significant environmental problems in the world. Therefore, this study proposes a carbon dioxide emission prediction model. Support vector machine and regression tree machine learning algorithms are adopted and compared. The proposed model is tested on a dataset collected from the Canada website throughout 2014 to 2020. The experimental results showed that the support vector machine learning algorithm is the most suitable algorithm for the adopted dataset. It obtained an overall 3.6026% RMSE, 12.978% MSE, 2.57% MAE, 100% R-Squared, almost 320 s in training time.

Keywords Carbon dioxide · Regression tree · Support vector machine · Prediction

1 Introduction

Carbon dioxide is one of the major components of Green House Gas (GHG) Emissions. It responsible for 81% from the total emissions [1]. It can be naturally produced from respiration or decomposition. Moreover, it can be produced from cement production, automotive exhausts, and the burning of fossil fuels. Methane, carbon dioxide, nitrous oxide, water vapors, and ozone are the main greenhouse gases that exist in the earth's atmosphere. However, all of these gases play a critical role to

G. I. Sayed (✉)
School of Computer Science, Canadian International College (CIC), New Cairo, Egypt
e-mail: Gehad_Sayed@cic-cairo.com
URL: https://www.egyptscience

A. E. Hassanien
Faculty of Computers and Artificial Intelligence, Cairo University,Giza, Egypt
URL: https://www.egyptscience

G. I. Sayed · A. E. Hassanien
Scientific Research Group in Egypt (SRGE), Giza, Egypt

keep life on the earth, and the concentration increase of any of those has a severe effect on climate change [2]. Greenhouse gases such as carbon dioxide emissions or shortly CO_2 are considered one of the essential causes of climate change. They can trap the temperature in the atmosphere. This can result in global temperatures rising. Thus it played a major role in global warming. Despite the natural sources of carbon dioxide such as volcanic eruptions, before the development of industry, humans are responsible for increasing carbon dioxide from 280 parts per million to 370 parts per million today [3]. Thus, a lot of labor and time is needed to prepare many prototypes and evaluate each one of them. Starting from the period 2000 to 2018, global greenhouse gas emissions of developed countries have decreased by 6.5%. Meanwhile, from 2000 to 2013, the emissions of developing countries increased to 43.2. This high increase is due to an increase in improved economic output evaluated in terms of GDP and industrialization [4].

1.1 Motivation

Car pollution is one of carbon dioxide sources. It can significantly affect water quality, air quality, and soil. It is one of the exhaust gas that has a dangerous effect on people suffering from heart disease and infants. This is due to its ability to interfere with the blood when delivering oxygen. Sulphur dioxide, formaldehyde, and benzene are other car pollutants that harm human health [5]. Thus, in the last few years, many researchers focus their studies to reduce pollutants and greenhouse gas emissions from vehicles.

Over the years, machine learning algorithms have proven their efficiency in the design, modeling, and prediction of the emissions of carbon dioxide. They showed their ability to find the approximate or optimal solution in many filed such as in agriculture [6], in medical [7], and in drug discovery [8]. Nowadays, many researchers used machine learning algorithms to increase the prediction accuracy of CO_2 emissions. Regression models are types of machine learning algorithms used for estimating the emission of CO_2 from cars. Authors in [5], used dynamic programming to predict the tank-to-wheel CO_2 emissions. They also proposed an automatic search tool to tune the hyper-parameters of the neural network. The experimental results showed that their proposed model obtained regression errors below 1%. In [9], the Gaussian Process Regression and the traditional parametric modeling algorithms are adopted and compared to predict CO_2 emissions. The simulation results revealed the efficiency and reliability of the proposed algorithms. In [10], the support vector machine (SVM) is used to predict the emission of carbon (CO_2) obtained from the Alcohol industry. The authors showed that the proposed model based on SVM obtained an overall 0.004 Root Mean Square Error (RMSE). Additionally, the results showed that the proposed model can be used for further monitoring expenditure of CO_2 emission. Authors in [1] proposed a hybrid approach based on using Long Short-Term Memory Network and Convolution Neural Network. they applied their proposed hybrid approach to predict the levels of CO_2 in India in 2020.

1.2 Main Contribution

This study proposes a carbon dioxide emission prediction model. The proposed prediction model consists of four main phases. The main objective of the first phase is to handle the missing values in the adopted dataset. Then the output from the previous phase is used to feed the prediction phase. Then, hold-out is used in the cross-validation phase to evaluate the reliability of the proposed prediction model. Finally, several measurements are used to evaluate the overall performance of the proposed prediction model with different machine learning algorithms. The main contribution of this study can be summarized as follows:

1. A carbon dioxide emission prediction model is proposed to predict the emission of CO_2 from a car.
2. Two well-known and commonly used machine learning algorithms known as support vector machine and regression tree are used and evaluated.
3. Several measurements are used to evaluate the proposed model. These measurements are R-squared, the mean squared error (MSE), the root mean squared error (RMSE), mean absolute error (MAE), and training time in seconds.

The rest of the work is organized as follows; Sect. 2 gives a detailed description of the proposed carbon dioxide emission prediction model. Section 3 provides the experimental results and discussions with a description of the used carbon dioxide dataset. Finally, conclusions and future research direction are presented in Sect. (4) .

2 The Proposed Carbon Dioxide Emission Prediction Model

The proposed carbon dioxide emission prediction model consists of four main phases; handling the missing values phase, the prediction phase, the cross-validation phase, and finally the evaluation phase. Overall the proposed carbon dioxide emission prediction model is shown in Fig. 1. In the first phase, all the missing values are replaced with the median value of that feature. Equation (1)shows the mathematical formula of the median method, where Y_i^t is the missing value for a given i-th iteration and d-th dimension and C_r is the median value of a class.

$$\overline{M}_d^i = median_{i:M_d^i \in C_r} M_d^i \tag{1}$$

In the prediction phase, two well-known and most used machine learning algorithms are used to predict the emission of carbon dioxide in a vehicle. These algorithms are regression trees with different types and the support vector machine. Support vector machine or shortly (SVM) is developed on statistical learning theory, where a group of mathematical functions namely known as kernels. Many studies showed its efficiency in classification, analysis, pattern recognition, and regression

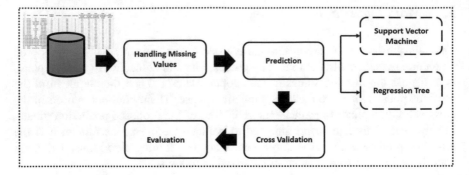

Fig. 1 The proposed carbon dioxide emission prediction model

tasks [10–12]. Also, the studies showed the capability of the SVM algorithm compared with the other algorithms. This is due to its ability to generalize. Thus, in this study, SVM with different kernel functions known as linear, Gaussian, cubic, and quadratic is adopted. Additionally, the regression tree with different splitting criteria is considered and evaluated in the proposed model. The regression tree is one of a multivariate, data-driven, nonparametric, and nonlinear machine learning algorithm. It is used to map the relationship between output data and input data algorithmically. The regression tree used this relationship and accumulate this learning into its constraints [13].

To evaluate the reliability and robustness of the proposed model, one of the cross-validation methods known as hold-out is used in the cross-validation phase. The data set in the holdout cross-validation method is divided into two sets namely; the testing set and the training set. A machine algorithm uses the training set only to build the training model. Then the testing set is used to predict the output values. In this work, the dataset is divided into 70% for training purposes and 30% for testing purposes. Finally, in the evaluation phase, several measurements are used. These measurements are R-squared, the mean squared error, the root mean squared error, mean absolute error, and training time in seconds.

3   Simulation Results and Discussions

In this section, different machine learning algorithms are evaluated and compared. All the conducted experiments are conducted on the same PC with specs. The detailed configuration settings are listed below:

- CPU: Intel Core i7
- OS: Windows 10
- RAM: 16 GB
- Language: Matlab R2020.

3.1 Performance Evaluation Measurements

Several measurements are used to evaluate how well a machine learning algorithm can predict the value of the response variable, which in our case is CO$_2$ emissions. These measurements are R-squared, the mean squared error (MSE), the root mean squared error (RMSE), mean absolute error (MAE), and training time in seconds. The mathematical formula of MSE is defined in Eq. (2), RMSE is defined in Eq. (3), and MAE is defined in Eq. (4), where y_i is the observed variable of the i-th observation, $\bar{y}_i$ is the predicted value of the i-th observation, N is the sample size. When the value of MSE, RMSE, and MAE is zeros, this means that the specified machine learning algorithm is the best algorithm that can accurately predict the observation value.

$$MSE = \sum \left(\frac{(\bar{y}_i - y_i)^2}{N} \right) \tag{2}$$

$$RMSE = \sqrt{\sum \left(\frac{(\bar{y}_i - y_i)^2}{N} \right)} \tag{3}$$

$$MAE = \sqrt{\sum \left(\frac{|(\bar{y}_i - y_i)|}{N} \right)} \tag{4}$$

R-squared is calculated by dividing the variance of the actual value of the response variable obtained by a regression machine learning algorithm rather than the MSE which captures the residual error. The mathematical definition is introduced in Eq. (5), where $\hat{y}_i$ is the mean value of the observed variable.

$$R^2 = 1 - \frac{\sum (\bar{y}_i - y_i)^2}{\sum (\hat{y}_i - y_i)^2} \tag{5}$$

3.2 Dataset Description

The adopted dataset has 7385 samples with total features equal to 11 [14]. This dataset is taken from the Canadian Government's official website [15]. It contains the fuel consumption and other describing features of a vehicle from 2014 to 2020. The description of these features is presented in Table 1. Figure 2 shows a sample of the used dataset.

Table 1 CO_2 emission dataset description

Feature	Description
Model	The data contain six car models; four-wheel drive (4WD/4X4), all-wheel drive (AWD), flexible-fuel vehicle (FFV), short wheelbase (SWB), long wheelbase (LWB), and extended wheelbase (EWB)
Fuel type	The fuel type of a car can be regular gasoline (x) or premium gasoline (z) or diesel (D) or ethanol (E) or natural gas (N)
Transmission	The transmission of a car can be automatic (A) or automated manual (AM) or automatic with select shift (AS) or continuously variable (AV) or manual (M)
Fuel consumption	Highway and city fuel consumption ratings are shown in litres per 100 km (L/100 km)—the combined rating (45% hwy, 55% city) is shown in L/100 km and in miles per imperial gallon (mpg)

1	Make	Model	Vehicle Class	Engine Size(L)	Cylinders	Transmission	Fuel Type	Fuel Consumpti on City (L/100 km)	Fuel Consumpti on Hwy (L/100 km)	Fuel Consumpti on Comb (L/100 km)	Fuel Consumpti on Comb (mpg)	CO2 Emissions(g/km)
2	ACURA	ILX	COMPACT	2	4	AS5	Z	9.9	6.7	8.5	33	196
3	ACURA	ILX	COMPACT	2.4	4	M6	Z	11.2	7.7	9.6	29	221
4	ACURA	ILX HYBRID	COMPACT	1.5	4	AV7	Z	6	5.8	5.9	48	136
5	ACURA	MDX 4WD	SUV - SMALL	3.5	6	AS6	Z	12.7	9.1	11.1	25	255
6	ACURA	RDX AWD	SUV - SMALL	3.5	6	AS6	Z	12.1	8.7	10.6	27	244
7	ACURA	RLX	MID-SIZE	3.5	6	AS6	Z	11.9	7.7	10	28	230
8	ACURA	TL	MID-SIZE	3.5	6	AS6	Z	11.8	8.1	10.1	28	232
9	ACURA	TL AWD	MID-SIZE	3.7	6	AS6	Z	12.8	9	11.1	25	255
10	ACURA	TL AWD	MID-SIZE	3.7	6	M6	Z	13.4	9.5	11.6	24	267
11	ACURA	TSX	COMPACT	2.4	4	AS5	Z	10.6	7.5	9.2	31	212
12	ACURA	TSX	COMPACT	2.4	4	M6	Z	11.2	8.1	9.8	29	225
13	ACURA	TSX	COMPACT	3.5	6	AS5	Z	12.1	8.3	10.4	27	239
14	ALFA ROMEC	4C	TWO-SEATER	1.8	4	AM6	Z	9.7	6.9	8.4	34	193
15	STON MARTI	DB9	MINICOMPACT	5.9	12	A6	Z	18	12.6	15.6	18	359
16	STON MARTI	RAPIDE	SUBCOMPACT	5.9	12	A6	Z	18	12.6	15.6	18	359
17	STON MARTN8 VANTAGE	TWO-SEATER	4.7	8	AM7	Z	17.4	11.3	14.7	19	338
18	STON MARTN8 VANTAGE	TWO-SEATER	4.7	8	M6	Z	18.1	12.2	15.4	18	354
19	STON MARTI8 VANTAGE	TWO-SEATER	4.7	8	AM7	Z	17.4	11.3	14.7	19	338
20	STON MARTI8 VANTAGE	TWO-SEATER	4.7	8	M6	Z	18.1	12.2	15.4	18	354
21	STON MARTI	VANQUISH	MINICOMPACT	5.9	12	A6	Z	18	12.6	15.6	18	359
22	AUDI	A4	COMPACT	2	4	AV8	Z	9.9	7.4	8.8	32	202
23	AUDI	A4 QUATTRC	COMPACT	2	4	AS8	Z	11.5	8.1	10	28	230

Fig. 2 Sample of the used dataset

3.3 Experimental Results

Figure 3 compares boosted ensemble tree (AdaBoost), and the bagged ensemble tree (Random Forest) with three regression trees with different values of the maximum number of divisions; (fine, medium, and coarse). It should be mentioned that the flexibility of a tree algorithm mainly depends on the maximum number of divisions. Additionally, the maximum number of divisions is used to control the depth of a tree. The maximum number of divisions in the fine regression tree is 100, while in the medium regression tree is 20, and in the coarse regression tree is 4. All of the used tree algorithms have the same splitting condition, where cross-entropy is used metric. As can be observed in Fig. 3, the fine regression tree obtained the lowest RMSE, MSE, and MAE, while the bagged ensemble tree namely Random Forest and medium regression tree are in the second place. Boosted ensemble tree namely AdaBoost obtained the worst results.

Figure 4 compares the R-Squared value of different regression trees. R-squared values range from zero to one. The one value of R-squared indicates the strong relationship between movements of a dependent variable based on an independent variable's movements. As it can be seen, the fine regression tree obtained the highest R-squared, which indicates that almost 100% of the variance of the dependent variable being studied is explained by the variance of the independent variable (CO$_2$ emissions). Moreover, it can be observed that boosted ensembles tree obtained the minimum R-squared. These results are consistent with the obtained results Fig. 3.

For further evaluation of the performance of the most popular regression trees, the computational time is used. Figure 5 compared the training time obtained from AdaBoost, random forest, and different settings of the regression tree. As can be observed, the regression tree with the medium maximum number of divisions obtained the minimum time. The fine regression tree is in third place after the coarse regression tree. The bagged ensemble obtained the maximum training time. From the previous experiments, it can be noted that the fine regression tree is the optimal regression tree for the CO$_2$ emissions dataset. As it can accurately predict CO$_2$ emissions for a vehicle in a reasonable time.

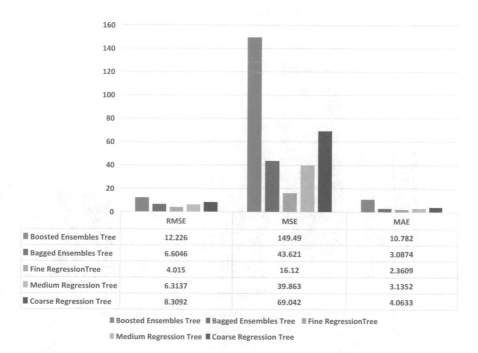

	RMSE	MSE	MAE
■ Boosted Ensembles Tree	12.226	149.49	10.782
■ Bagged Ensembles Tree	6.6046	43.621	3.0874
▨ Fine RegressionTree	4.015	16.12	2.3609
▨ Medium Regression Tree	6.3137	39.863	3.1352
■ Coarse Regression Tree	8.3092	69.042	4.0633

■ Boosted Ensembles Tree ■ Bagged Ensembles Tree ▨ Fine RegressionTree
▨ Medium Regression Tree ■ Coarse Regression Tree

Fig. 3 Boosted ensembles tree versus bagged ensembles tree versus fine regression tree versus medium regression tree versus coarse regression tree in terms of RMSE, MSE, and MAE

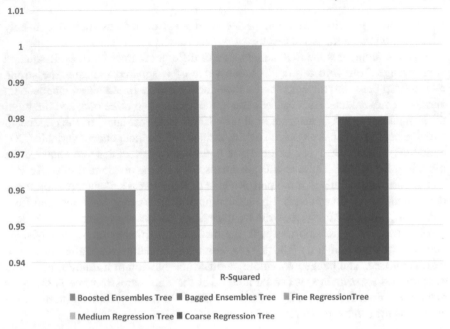

Fig. 4 Boosted ensembles tree versus bagged ensembles tree versus fine regression tree versus medium regression tree versus coarse regression tree in terms of R-squared

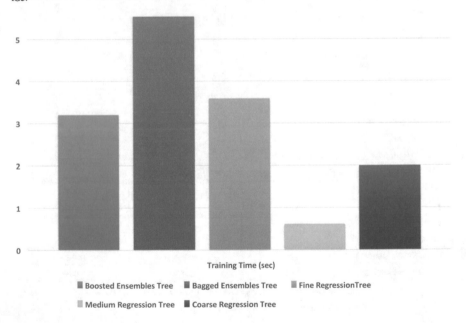

Fig. 5 Boosted ensembles tree versus bagged ensembles tree versus fine regression tree versus medium regression tree versus coarse regression tree in terms of training time in seconds

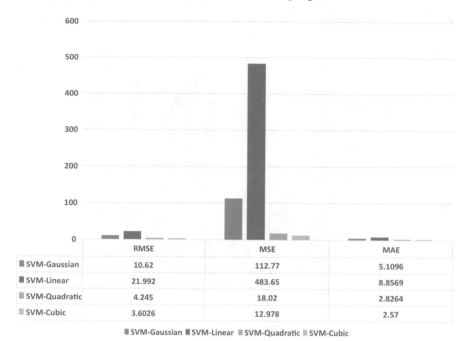

	RMSE	MSE	MAE
■ SVM-Gaussian	10.62	112.77	5.1096
■ SVM-Linear	21.992	483.65	8.8569
■ SVM-Quadratic	4.245	18.02	2.8264
■ SVM-Cubic	3.6026	12.978	2.57

■ SVM-Gaussian ■ SVM-Linear ■ SVM-Quadratic ■ SVM-Cubic

Fig. 6 SVM With Gaussian kernel versus SVM with linear kernel versus SVM with quadratic kernel versus SVM with cubic kernel in terms of RMSE, MSE, and MAE

In the next experiments, one of the most popular machine learning algorithms called support vector machine is adopted. Figure 6 compares RMSE, MSE, and MAE obtained from the support vector machine (SVM) regression algorithm with different kernel functions; linear, Gaussian, Quadratic, and Cubic. As it can be observed, SVM with cubic kernel function is the optimal kernel function that can significantly predict the value of CO$_2$ emission. SVM with quadratic kernel function is in the second place, while SVM with linear kernel function is the worst one. This is due to the adopted dataset being complex that can't be easily separated using a simple linear function.

Figure 7 compares the R-squared value of SVM with Gaussian kernel, SVM with Linear Kernel, SVM with Quadratic Kernel, and SVM with Cubic Kernel. As it can be seen, SVM with Cubic Kernel achieved the highest R-squared, while SVM with Linear Kernel obtained the lowest R-Squared. This result is consistent with the obtained result from the previous experiment.

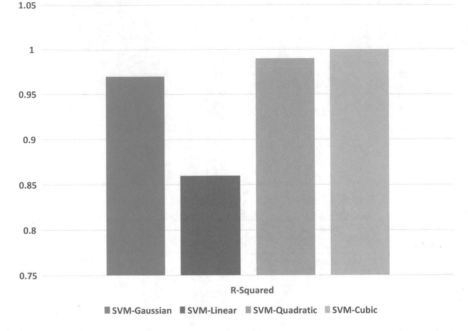

Fig. 7 SVM with Gaussian kernel versus SVM with linear kernel versus SVM with quadratic kernel versus SVM with cubic kernel in terms of R-squared

Figure 8 compares the training time in seconds for the support vector machine with four kernel functions, cubic, Gaussian, linear, and quadratic. As it can be observed, SVM with Gaussian kernel function obtained the less training time compared with the other kernel functions, while the SVM with cubic kernel function registered the highest training time.

Figure 9 shows predicted CO_2 emissions from SVM with cubic kernel function with the actual CO_2 emissions. As it can be observed, the predicted values of CO_2 emission are almost identical to the actual values of CO_2 emissions. From all the obtained results, it can be concluded that SVM with cubic kernel function is the most suitable machine learning algorithm for predicting the values of CO_2 emissions obtained from different models of cars. It obtained overall 3.6026% RMSE, 12.978% MSE, 2.57% MAE, 100% R-Squared, almost 320 s in training time, while, the fine regression tree obtained overall 4.015% RMSE, 16.12% MSE, 2.3609% MAE, 100% R-Squared, almost 3.5 s in training time.

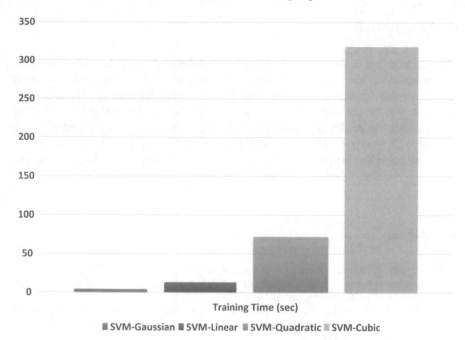

Fig. 8 SVM with Gaussian kernel versus SVM with linear kernel versus SVM with quadratic kernel versus SVM with cubic kernel in terms of training time in seconds

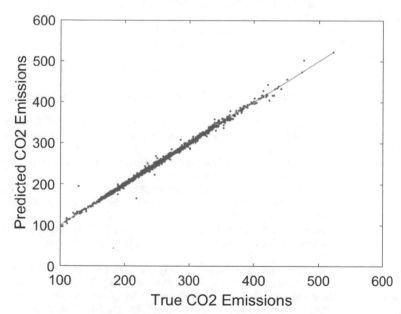

Fig. 9 Predicted versus actual plot using SVM with cubic kernel

4 Conclusions

The normal range of the temperature is considered one of the significant reasons for the sustenance of life on the earth. carbon dioxide is one of the atmospheric gases that are responsible for 81% of the total emissions. Additionally, carbon dioxide shortly CO_2 has played a critical role in global warming. This work introduces a carbon dioxide emission prediction model to predict CO_2 emission. The proposed model is applied to CO_2 emissions from cars in Canada from 2014 to 2020. The experimental results revealed that the proposed model can effectively be used as a decision making for monitoring the emission of CO_2 from the cars.

References

1. Amarpuri, L., Yadav, N., Kumar, G., & Agrawal, S. (2019). Prediction of CO_2 emissions using deep learning hybrid approach: a case study in Indian context. In *2019 twelfth international conference on contemporary computing (IC3)* (pp. 1–6).
2. Saleh, C., Leuveano, R., Ab Rahman, M., Deros, B., & Dzakiyullah, N. (2015). Prediction of CO_2 emissions using an artificial neural network: The case of the sugar industry. *Advanced Science Letters, 21*, 3079–3083.
3. Li, S., Siu, Y., & Zhao, G. (2021). Driving factors of co2 emissions: Further study based on machine learning. *Frontiers in Environmental Science, 9.*
4. Deepthi, A., Divya, A., Rajini, D., Khan, G., & Inampudi, L. (2021). System to predict CO_2 emission levels in cars to mitigate air pollution, *13*, 521–527.
5. Maino, C., Misul, D., Di Mauro, A., & Spessa, E. (2021). A deep neural network based model for the prediction of hybrid electric vehicles carbon dioxide emissions. *Energy and AI, 5*, 100073.
6. Sayed, G., Hassanien, A., & Tang, M. (2022). A novel optimized convolutional neural network based on marine predators algorithm for citrus fruit quality classification. In X. Shi, G. Bohács, Y. Ma, D. Gong, & X. Shang (Eds.), *LISS 2021* (pp. 682–692). Singapore: Springer Nature Singapore.
7. Sayed, G., Soliman, M., & Hassanien, A. (2021). A novel melanoma prediction model for imbalanced data using optimized squeezenet by bald eagle search optimization. *Computers in Biology and Medicine, 136*, 104712.
8. Gupta, R. (2022). Application of artificial intelligence and machine learning in drug discovery. *Methods in Molecular Biology, 2390*, 113–124.
9. Ma, N., Shum, W., Han, T., & Lai, F. (2021). Can machine learning be applied to carbon emissions analysis: An application to the CO_2 emissions analysis using Gaussian process regression. *Frontiers in Energy Research, 9.*
10. Saleh, C., Dzakiyullah, N., & Nugroho, J. (2016). Carbon dioxide emission prediction using support vector machine. In *IOP conference series materials science and engineering, 114*, 012148.
11. Rao, M. (2021). Machine learning in estimating CO_2 emissions from electricity generation. In M. Sales Guerra Tsuzuki & R. O. O. Abdel Rahman (Eds.), *Engineering Problems—Uncertainties, Constraints and Optimization Techniques*, Chap. 1. IntechOpen, Rijeka.
12. Mladenović, I., Sokolov-Mladenović, S., Milovančević, M., Marković, D., & Simeunović, N. (2016). Management and estimation of thermal comfort, carbon dioxide emission and economic growth by support vector machine. *Renewable and Sustainable Energy Reviews, 64*, 466–476.

13. Xiwen, C., Shaojun, E., Dongxiao, N., Bosong, C., & Jiaqi, F. (2021). Forecasting of carbon emission in China based on gradient boosting decision tree optimized by modified whale optimization algorithm. *Sustainability, 13*(21), 1–18.
14. Canada government official website. https://open.canada.ca/data/en/dataset/98f1a129-f628-4ce4-b24d-6f16bf24dd64#wb-auto-6
15. Podder, D. Basic EDA of the CO$_2$ emissions by vehicle dataset. https://www.kaggle.com/code/debajyotipodder/basic-eda-of-the-co2-emissions-by-vehicle-dataset/notebook

Climate Change: The Challenge of Tunisia and Previsions for Renewable Energy Production

Wahiba Ben Abdessalem⬤, Ilyes Jayari, and Sami Karaa

1 Introduction

Climate change relates to global warming, sea level rise, changes in storms and monsoons, drought, and melting permafrost [1, 2]. Climate damage will lead to increased inequality because raised impacts can be expected, especially in warmer regions, which are often linked to poorer countries, including the Middle East and North Africa (MENA) [3].

Based on an analysis carried out by the International Renewable Energy Agency (IRENA), which is an intergovernmental organization supporting countries in their transition to sustainable energy [4], energy-related carbon dioxide (CO_2) emissions would need to be reduced by about 70% by 2050, compared to current levels. The extensive use of electricity from renewable energies could help reduce CO_2 by 60%, or even 75%, if renewable energies are used for heating and transport.

According to the report, the global demand for electrical energy continues to increase. Renewable energies, such as solar and wind, could meet 86% of electricity demand [4].

Aware of these threats, Tunisia has adopted a proactive policy to combat climate change. Tunisia submitted its Intended Nationally Determined Contribution (INDC)

W. Ben Abdessalem (✉)
High Institute of Management of Tunis, Tunis University, Tunis, Tunisia
e-mail: wahiba.abdessalem@isg.rnu.tn

RIADI-GDL Laboratory, ENSI, University of Manouba, Manouba, Tunisia

I. Jayari
Tunis, Tunisia
e-mail: jayarilyes@gmail.com

S. Karaa
Société Tunisienne de L'Electricité et du Gaz (STEG), Tunis, Tunisia
e-mail: skaraa@steg.com.tn

© The Author(s), under exclusive license to Springer Nature Switzerland AG 2023
A. E. Hassanien and A. Darwish (eds.), *The Power of Data: Driving Climate Change with Data Science and Artificial Intelligence Innovations*, Studies in Big Data 118,
https://doi.org/10.1007/978-3-031-22456-0_6

to the Conference of Parties of the United Nations Framework Convention on Climate Change (UNFCCC) on September 16, 2015. [5]

Tunisia proposes to reduce its greenhouse gas emissions in all sectors (energy, industrial processes, agriculture, forest, and other land uses, and waste) to reduce its carbon intensity by 46% in 2030 compared to the base year 2010. The production of electrical energy represents the largest sector of CO_2 emissions. Consequently, Tunisia has focused primarily on this sector, which alone could contribute 75% of emission reductions.

The energy mix represents the solution. Several studies on the energy mix for electricity production in Tunisia have been carried out. The study by Lechtenböhmer et al. in 2012 focused on modeling and analyzing several scenarios from 2009 to 2030. This study has shown that none of the scenarios studied successfully reduces the demand for non-renewable energy and related greenhouse gas emissions. Renewable energies are the only scenario that can mitigate them [6].

This chapter is divided into seven sections. The following Sect. 2 represents the study area. Section 3 explains the energy situation in Tunisia. Section 4 explains the Tunisian commitments to climate change in the energy sector. Section 5 summarizes the results of the study carried out to establish the inventory of solar photovoltaic (PV) projects currently connected to the Low Voltage (LV), Medium Voltage (MV), and High Voltage (HV) network in self-production mode as well as the evaluation of the potential for PV self-production and forecasts of evolution by 2030. Section 6 proposes a decision support framework for PV energy prediction. A conclusion and discussions are presented at the end of the chapter.

2 Study Area

Tunisia is located in the North of Africa (refer to Fig. 1). It is a bath by the Mediterranean Sea to the north and east, bordered by Libya to the South and most south, and Algeria to the southwest and west.

Tunisia is divided into two large geographical areas:

- A northwestern zone with chaotic reliefs delimiting a series of high plains
- A southeastern zone with a low and hilly appearance extending to the coast.

The relatively high latitude of Tunisia and its stretch geographically from north to the South give it the succession of climatic zones ranging from sub-humid to the north, semi-arid to arid in Tunisia's central, to desert for all the South finally.

With a climate marked by aridity, Tunisia is considered among the Mediterranean countries most exposed to climate change, with the risk of a sharp increase in temperature. This increase would vary by the region, with the best case between 1 and 1.8 °C by 2050 and between 2 °C and 3 °C at the end of the century. In the pessimistic case, the increase could reach 4.1 to 5.2 °C at the end of the century. Projections show a decline in precipitation (-10 to 30% in 2050), a rise in the level of the sea (30–50 cm in 2050), and other phenomena of climatic extremes (floods and droughts) [7].

Fig. 1 Geographic location of Tunisia

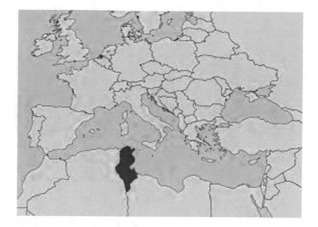

These climatic risks would have adverse effects on the social, economic, and ecological, which would manifest themselves in the scarcity of water resources, the weakening of ecosystems land, and sea, the decline in agricultural activities and tourism, and the strengthening of the capitalization of economic activities.

The new Constitution of Tunisia considered the development sustainability and the rights of future generations among the fundamental rights of Tunisians.

The joint responsibility of the State and society in the preservation of water resources, the fight against climate change, and the right to a healthy environment for all citizens has been enshrined as a priority in article 45 of the constitution.

Tunisia supports the United Nations Development Program (UNDP) [8], a United Nations agency for international development, working in 170 countries to eradicate poverty, decrease inequalities, and assist countries in developing policies, leadership skills, and partnership capacities. UNDP support to Tunisia focuses on three key areas: democratic governance and consolidation of reforms; inclusive growth and sustainable human development, the environment; and the fight against climate change.

3   The Energy Situation in Tunisia

The energy sector plays a key role in the success of all policies, as do the economic and social sectors. It is also of great strategic importance, especially in light of the climate changes taking place in the world. Tunisia is facing strategic, environmental, societal, and economic challenges.

During 2010–2021, the resources available in primary energy in Tunisia stand at approximately 5.1 million tonnes of oil equivalent (toe). The energy mix is currently dependent on 53% natural gas and 47% petroleum materials, while the contribution of renewable energies does not exceed 0.4% [9].

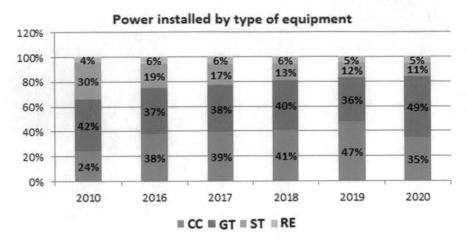

Fig. 2 Electricity production in Tunisia [6]

The electricity production fleet is divided into 27 units, using: Gas Turbines (GT), Steam Turbines (ST), Combined Cycles (CC), and Renewable Energies (RE) between 2010 and 2020. Installed capacities are currently distributed as exposed in Fig. 2.

4 Energy and Climate Change

Tunisia has adopted a proactive policy to fight against climate change both in terms of mitigation and adaptation. The international negotiations on climate change organized within the United Nations Framework Convention on Climate Change [10] led to a historic agreement in December 2015 in Paris called the "Paris Agreement" [11]. This agreement invited all the countries party to the UNFCCC to adopt public policies to contain the increase in temperatures below 2 °C or even 1.5 °C by 2100. To achieve this objective, all the parties are called upon to establish, communicate and update their Nationally Determined Contributions (NDCs) every five years. The NDC represents the political instrument that officially translates the commitments of each country to contribute to the international effort to fight against climate change.

Tunisia submitted its first NDC in September 2015, the objective of which is to reduce the carbon intensity of all sectors of the economy by 40% by 2030 compared to its level in 2010. Energy is placed at the heart of the priority sectors in the mitigation field, with a substantial contribution of 75% to the overall mitigation objective of the Tunisian NDC. Energy efficiency and renewable energies are the two main levers for achieving the objective assigned to the energy sector.

The mitigation effort will come more particularly from the energy sector, which accounts for 75% of emission reductions. It is expected that the energy sector will reduce its carbon intensity in 2030 by 46% compared to 2010 (Fig. 3) [7, 12], as part of the energy transition policy recommended by the State. Despite the efforts made

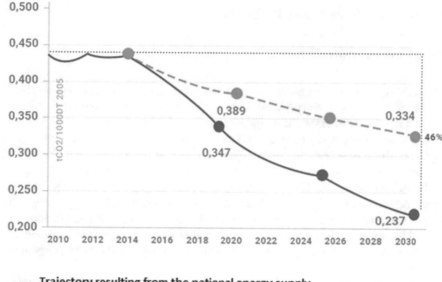

—— Trajectory resulting from the national energy supply
— — Energy Mitigation Goal Trajectory

Fig. 3 Decrease of carbon intensity [7]

by Tunisia for three decades in terms of energy control and to meet these challenges, the Tunisian authorities have decided since 2013 to engage in an unprecedented strengthening of the energy control policy. Energy with its two components of energy efficiency and renewable energies. This transition targets, by 2030, a reduction in primary energy demand of 30% compared to the trend scenario and a penetration rate of renewable energies in electricity production of 30%.

However, energy challenges persist; demand remains dominated by hydrocarbons (natural gas and petroleum products) which cover 99% of primary energy consumption, while renewable energies (excluding biomass) do not exceed 1% of this consumption. Because of this increased dependence on conventional energies, coupled with the drop in national hydrocarbon production, the equilibrium of the energy balance was broken from the beginning of 2000. In the future and based on current conventional resources, the energy forecast shows significant challenges regarding the security of the country's energy supply. Indeed, if energy demand evolves according to a trend scenario, the energy balance deficit would reach around 13.3 Mtoe in 2030. In the case of an energy efficiency scenario, this deficit would be 7.9 Mtoe.

5 Renewable Energies in Tunisia

5.1 Current Situation

Tunisia's non-renewable resources are modest compared to international standards and benchmarks; the country. On the other hand, it has strong wind potential and solar resources, which are among the highest in the world. In addition, several resource assessments have already been carried out in Tunisia, which places the country in a good position to accelerate the deployment of renewable energy technologies and translates, concretely, into a real lead over many other countries in the MENA region.

The wind resource has been assessed by the National Agency for Energy Management (Agence Nationale pour la Maitrise de l'Energie: ANME) [13] through the development of a wind atlas for the whole of Tunisia as part of cooperation with Spain [14]. Indeed, the Wind Atlas indicates that the wind conditions are suitable and show a very interesting potential in several regions of Tunisia, particularly in the South.

Tunisia also benefits from a significant rate of sunshine exceeding 3000 h per year. Tunisia is a small country, barely 750 km long (from north to South). However, the sunshine varies significantly depending on the region. The direct irradiation index varies from 1800 kWh/m^2.year in the far north to 2600 kWh/m^2.year in the far South.

As part of the solar market strengthening project in Tunisia, the German international development cooperation agency GIZ (Deutsche Gesellschaft für Internationale Zusammenarbeit) [15] carried out a study in the context of a cooperation project for Strengthening the Solar Market in Tunisia [16]. The Project provides technical support and advice and an exchange of international experiences to develop a relevant, incentive, and operational regulatory framework.

The Tunisian government is successfully implementing the Tunisian Solar Plan (TSP) [17], developing large-scale renewable energies and respecting the country's agreed contributions to climate protection. Figure 4 shows the global objectives of the Tunisian solar plan [17].

To implement these ambitious objectives, a new law on the production of electricity from renewable energies was promulgated in 2015 (n°2015-12 of May 11, 2015) thus determining the different production regimes and authorizing the private sector to play a greater role in achieving the objectives set by the State, through the following production schemes:

– Concession regime for projects whose power exceeds 10 MW for solar energy, 30 MW for wind energy, 15 MW for biomass, and 5 MW for other forms.
– System of authorizations for projects whose power does not exceed the above-indicated thresholds.
– Self-production regime for any type of customer.

Self-production is a major axis of the national energy transition policy contributing to the energy mix's diversification. It is dedicated to playing an important role in achieving objectives in terms of electricity production from renewable energies.

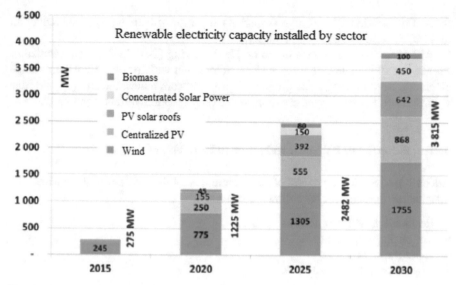

Fig. 4 Objectives of the Tunisian solar plan [17]

5.2 Self-Production of Photovoltaic Solar Energy

The self-production of electricity remains possible for any local authority and a public or private establishment, connected to the national electricity grid at Medium or High Voltage (MV-HV). It concerns establishments active in the industrial, agricultural, or tertiary sectors. It is possible to submit a request to the Ministry of Energy, Mines, and Energy Transition [Ministère de l'Energie des Mines et des Energies Renouvelables [18] to install the necessary equipment for the MV/HT self-production of electricity which will be sanctioned by an agreement [19].

The self-production program allows the deployment of two types of projects:

– *On-site projects*, without transmission of electricity on the national network of the Tunisian Electricity and Gas Company (Société Tunisienne de l'Electricité et du Gaz (:STEG) [20].
– *Projects on a remote site*, with the transport of electricity on the STEG network.

The electrical energy transferred by the self-producer to STEG as surplus production from the renewable installation is shifted and then invoiced monthly. The contractual relations between the self-producer and STEG are defined in a contract.

Among all renewable technologies, solar photovoltaic technologies have dominated the renewable energy industry worldwide for many years. Photovoltaic solar installations could multiply by six over the next ten years and reach an annual increase of 9% until 2050. By 2050, the self-production of PV origin will represent 40% of the total capacity projected [21].

The International Energy Agency (IEA) [22] deployed a technology roadmap for PV energy. The roadmap assumes that the costs of electricity from PV in different parts of the world will converge as markets develop, with an average cost reduction of 45% by 2030 and 65% by 2050 [23].

With the metering system based on Net-Metering [24], the Tunisian regulatory framework allows subscribers to the Low Voltage network to cover all of their annual electricity needs through self-production projects from renewable energies. For this voltage level, these energies are represented only by photovoltaic technology.

5.2.1 Potential of PV Self-Production in Low Voltage

According to STEG, at the end of 2019, there were 4,049,047 subscribers with a PV installation, spread over the following sectors:

- Residential, which represents 86.3%.
- The tertiary represents 10.7%.
- Agriculture with 2%.
- The industrial with only 1%.

Currently, the PV market is facing significant growth in demand. This is because STEG has difficulty developing enough capacity to cover national demand. The advantage of developing the LV market is that private households have the opportunity to contribute financially to a long-term energy transition. All the power installed on the PV market in LV would be financed exclusively by private investment.

The study conducted in this context aims to make estimates and show the potential for the years to come. Regarding the development potential of the residential sector, the study focused on the availability of surface, the technical potential of total power to be installed, and the available funding programs:

A—Availability of surface

This involves estimating the surface of the roofs available for the installations, based on studies by GIZ [25] and the National Institute of Statistics on the characteristics of housing [26]. It is estimated that 40% of the total roof surface is available for photovoltaic systems and about 60% of the remaining surface for the use of satellite dishes, and other uses by the inhabitants. Given these estimates, the estimated available surface for the residential sector does not present any problem with the availability of roof surfaces.

B—Technical potential of total power to be installed

At the end of 2018, approximately 62,000 residential subscribers opted for a self-production installation from renewable energies. Based on data from STEG's 2018 annual report [27]. Taking into account the total consumption of this segment at the end of 2019, and with an annual growth of 3%, this potential will increase to around 4000 MWp by the year 2030.

C—Available funding programs

ANME has implemented, since 2018, the social photovoltaic program "Social PV" for households with low electricity consumption (less than 1800 kWh per year), fully financed by the Energy Transition Fund (FTE) up to 15MDT. The program promotes the generalization of the installation of photovoltaic solar panels in households.

Also, an Economic PROSOL Program has been set up for 2019–2023. This program is part of the project "Promotion of renewable energy and energy efficiency in the building sector in Tunisia", funded by the NAMA Facility [28]. With an estimated budget of €5.3 million, this fund was granted as a donation. This NAMA Facility program aims to encourage households whose electricity consumption is less than 1800 KWh per year.

Regarding the development potential of the non-residential sector, it should be noted that non-residential subscribers generally have higher than average monthly electricity bills.

Aiming to involve public establishments in the energy transition and in achieving the objectives of the Tunisian Solar Plan in terms of renewable energies, ANME and the Ministry of Industry and Small and Medium-Sized Enterprises jointly developed the 2017 program to equip public buildings with photovoltaic installations under the self-production scheme in collaboration with the German Development Bank KfW for seven years. This Public program, called PROSOL Public Program, which is scheduled to be implemented in early 2021, will only focus on promoting the use of photovoltaics for producing electricity in the public sector, given the significant potential it offers. It provides for installing a hundred photovoltaic systems with an approximate capacity of 30 Megawatts by 2024.

5.2.2 Potential of PV Self-Production in Medium and High Voltage

This market is divided into three types of activity: tertiary, industry, and farming. For this voltage category, Tunisian regulations have limited the excess electricity from renewable energy self-production facilities that could be transferred to STEG. Decree No. 2016-1123 requires that the surpluses sold to STEG do not exceed 30% of the annual production of the self-production facility. Thus, the quantities of electricity exceeding this limit will be transferred free of charge to STEG. The total number of MV/HV customers is 19,701 [27]. The distribution of electricity consumption by sector, based on the STEG annual report for 2018 is presented in the following table (Table 1).

Table 1 Distribution of electricity consumption by branch [27]

	Sector	Consumption (GWh)	% Company
Industry	Extractive industry	308	3.78
	Food industry	687	8.42
	Textile industry	457	5.60
	Paper industry	110	1.35
	Chemical industry	493	6.04
	Construction materials industry	1472	18.04
	Metallurgical industry	284	3.48
	Miscellaneous industry	1159	14.21
Others	Agriculture	606	7.43
	Pumping	766	9.39
	Transportation	305	3.74
	Tourism	540	6.62
	Service & others	971	11.90
	Total	8158	100

6 Decision Support Framework for Photovoltaic Energy Prediction

According to an analysis of the inventory of photovoltaic solar projects under the self-production regime, we have noted a weak adhesion of individuals, institutions, and Tunisian companies connected to the Medium Voltage and High Voltage electricity networks. This is due to various regulatory, institutional, technical, and economic constraints.

It is in this context that we propose a framework taking into account the inventory of PV solar photovoltaic projects connected to the medium voltage network authorized under the self-production regime to monitor and provide the necessary information to decision-makers for the forecasts and decision-making regarding a greater development of achievements under this regime. This framework is a decision-making system based on a data warehouse. To do this, we started with data collection to be able to build the data warehouse.

6.1 Data Collection

In this study, the data is collected from energy consumption databases, and other data is collected in Excel sheets. Also, a self-producer (or self-consumer) survey is performed.

The survey was carried out among all owners of PV projects connected to the authorized MV network and their installers. The objective is to draw up an inventory of the progress of the projects of Medium Voltage (MV) self-consumption to:

– Determine the installed power distribution by region and sector of activity.
– Identify the progress of PV installations (rate of realization of MV projects published in the Journal Officiel de la République Tunisienne (Journal Officiel de la République Tunisienne: JORT), total power installed and connected to the MV network).
– Assess the difficulties encountered in implementing MV projects.
– Identify the rate of satisfaction of the beneficiaries.

As part of this study, the questionnaire developed for the telephone and on-site survey meets the following requirements:

– The questionnaire must be adapted to the type of activity;
– The form of the questions must make it possible to collect the desired information;
– The terms used in the questionnaire must be easily understandable and unambiguous.

The developed questionnaire addresses the descriptive and indicative elements of the status of the MV projects defined in the database, including in particular:

– Customer characteristics (tariff regime before/after installation).
– Installed power.
– Description of the installation (Types, power, number, and origin of modules and inverters as well as the installation method).
– The investment cost, the amount of the subsidy and its payment.
– The method of financing and facilities, the share of self-financing as well as profitability indicators.
– Performance of the facilities (ratios of self-consumption / coverage of needs/surplus).

Also, the questionnaire details the following aspects:

– The deadlines for implementing MV projects (administrative procedures and on-site construction).
– The management of surplus sales and invoicing by the beneficiary of systems (The monitoring and control carried out by the beneficiary and the invoicing method of STEG).
– The equipment guarantees and their receptions, the operation and maintenance of the PV system, and communication with the installing company (Contract, others).

- Energy efficiency measures are envisaged by the beneficiary.
- Comparison between the electricity bill before and after the implementation of the PV installation.
- Customer satisfaction about the savings made and the difficulties/advantages of the assembly process and implementation of an MT project in general.
- A brief visual inspection of the installation and, if necessary, some pictures.
- Self-producer recommendations.

Thus, the telephone survey affected 118 projects out of the 156 authorized. Following the results of the telephone survey, a representative sample of 12 self-consumers and 18 projects representing more than 10% of all authorized projects was chosen to conduct face-to-face interviews with the beneficiaries of these projects.

Finally, the data collected from the various sources relate in particular to:

- Company name or name of the self-consumer, as well as their contact details.
- Contact person.
- Site coordinates.
- Governorate and Delegation.
- Sector and sub-sector of activity.
- STEG reference, type of tariff, and district concerned by the reception and connection of the PV installation.
- Installing company, as well as its contact details.
- System size (Unit power in kilowatt peak (kWp)).
- Percentage of consumption coverage.
- JORT number, and the date of allocation of the authorization.
- Installation date.
- Date of approval of the detailed study, the signature of the STEG contract, date of commissioning.
- Information concerning the funding.
- The producible of the last 3 years, etc.

6.2   The Proposed Data Warehouse

We propose an approach based on Data Warehouse (DW) technology to address the issues related to PV energy provisions. The latter will support the analysis of PV electricity data to deliver reports and useful information for decision-makers. Therefore, the DW-based approach is introduced to manage and analyze PV energy consumption and production data, delivering valued information for decision-making.

A data warehouse is a copy of transactional data specifically structured for querying and analysis [29]. A DW is a decision support database that is maintained separately from the organization's operational databases. A data warehouse is a Subject-oriented, integrated, time-varying, non-volatile, collection of data used primarily in organizational decision-making [30].

A data warehouse is constructed by integrating data from multiple heterogeneous sources that support analytical reporting, structured and/or ad hoc queries and decision making [31].

DW includes 4 main components: data sources, ETL, data warehouse, and data Access and Analysis [32]:

- Data Sources, gathering the input data as raw material for the data warehouse, including operational databases, and data files (excel, CSV...).
- ETL (Extract, Transform, Load) process, is needed to extract data from data sources, transform the data for integration needs, and load the transformed data into the DW. Several ETL tools exist, such as TALEND, PENTAHO, CLOVERETL, etc. Selecting the right ETL tool is a crucial task for any DW. Each tool has its advantages and disadvantages [33].
- Data warehouse, which can be composed of data marts to store the loaded data in an organized way. Before choosing the final tool to implement the DW (SQL Server, Informix, Hyperion, ...), ensure that the tool is capable of meeting the growth and overall requirements of the organization in the present and the future [34].
- Data Access and Analysis, this component is used by decision-makers to access the DW for analysis target. It helps to derive insights from data to be able to make strategic decisions. Several Business Intelligence tools, such as Oracle BI, Microsoft Power BI, SAS Business Intelligence, etc., make it possible to achieve these objectives [35].

Three models are used to design a Data warehouse: the star, snowflake, and galaxy model. The star and snowflake models are the most used in companies. The difference between star and snowflake models is that the star model does not use normalization, whereas the snowflake model uses normalization to eliminate data redundancy. The two main components of these models are dimensions and facts.

- The dimensions are the axes we want to carry out the analysis. There may be dimension hierarchies to split dimension tables when they are too large. A dimension is a table with a primary key and a list of attributes. A dimension table must be linked to a fact table.
- The fact tables are those on which the analysis will focus. These tables contain operational information and relate to the life of a company. The fact table helps the user analyze the business dimensions, which helps in making decisions to improve their business. The fact table contains a primary key which is a concatenation of primary keys of all dimension tables, and numerical variables called measures which can be aggregated (SUM, AVG, COUNT...) using the attributes of the dimension tables.

In this study, the data warehouse incorporates data from the data found in energy consumption databases, the questionnaire, and the data saved in Excel sheets. Table 2 convenes these data.

The conceptual model of the proposed data warehouse according to the snowflake model is described in Fig. 5.

Table 2 Dimension and fact tables

	Tables	Attributes
Dimensions	Self-producer	IdSP (primary key), name, address, energy consumption
	Region	IdR (primary key), name
	Governorate	IdG (primary key), name, IdR (foreign key)
	Delegation	IdDel (primary key), name, IdG (foreign key)
	Installation company	IdIS (primary key), name
	STEG reference	IdSR (primary key), district name
	Tariff type	IdTT (primary key), tariff type
	Installation site	IdIS (primary key), name
	Main activity sector	IdMA (primary key), name
	Secondary activity sector	IdSA (primary key), name, IdMA (foreign key)
	Date	IdD (primary key), day, month, year
	Financing type	IdFT (primary key), financing type
Fact	Production PV	IdSP, IdG, IdDel, IdIS, IdSR, IdTT, IdIS, IdMA, IdSA, IdD, IdFT (foreing keys)
		Satisfaction rate
		Advancement rate
		Number of installations
		Installed power
		Energy produced
		Energy injected into the steg network

Once implemented, the DW allows managers to:

- Display the number of installations by date of commissioning.
- Have the coverage percentage (PV consumption/production).
- Select data relating to such period, production, sef-producer sector, etc.
- Sort group, or distribute this data according to the criteria of their choice.
- Perform calculations (totals, averages, differences, comparisons from one period to another, etc.).
- Present the results in a synthetic or detailed way, with a graph according to their needs or the expectations of the decision-makers.

7 Conclusions and Recommendations

Tunisia's energy situation is marked by limited energy resources, a decline in energy production, and a strong increase in demand. To follow the path of sustainable development in Tunisia, it is imperative to develop renewable energies and massively accelerate of energy efficiency projects.

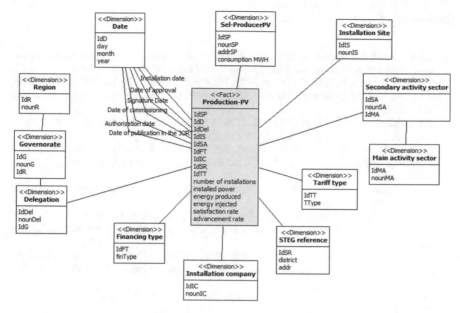

Fig. 5 Proposed Data Warehouse

For sustainable development and to mitigate the impact of global climate change, Tunisia is committed to reducing greenhouse gases. The effort has focused on introducing renewable energies to partly replace conventional energy production, the main emitter of CO_2. In this work, a study is carried out to establish the inventory of renewable energy projects, particularly the solar photovoltaic currently connected in self-production mode. An evaluation of the potential for PV self-production and forecasts of evolution by 2030 is also performed.

For the achievement of the objectives established by the government and to accelerate the implementation of renewable energy installations such as PV self-production projects, we recommend a new strategy based on four main elements:

- Respect regulatory deadlines.
- Change the regulations to simplify the procedure as much as possible.
- Allow some flexibility (like reprogramming counters).
- Change the counting mode: from the instantaneous to the hourly post with a monthly balance sheet (change at the level of the contract).

References

1. Lionello, P., & Scarascia, L. (2018). The relation between climate change in the Mediterranean region and global warming. *Regional Environmental Change, 18*(5), 1481–1493.
2. Masson-Delmotte, V., Zhai, P., Pörtner, H. O., Roberts, D., Skea, J., Shukla, P. R., & Waterfield, T. (2018). Global warming of 1.5 C. *An IPCC Special Report on the impacts of global warming of 1.5 °C, 1*(5).
3. Burke, M., Hsiang, S. M., & Miguel, E. (2015). Global non-linear effect of temperature on economic production. *Nature, 527*(7577), 235–239.
4. IRENA. (2019). *Global energy transformation: A roadmap to 2050* (2019 edition). https://www.irena.org/publications/2019/Apr/Global-energy-transformation-A-roadmap-to-2050-2019Edition
5. Conference of the Parties (COP). (2022). https://unfccc.int/process/bodies/supreme-bodies/conference-of-the-parties-cop
6. Lechtenböhmer, S., Durand, A., Fischedick, M., Nebel, A., Soukup, O., Wane, S., & Zarrouk, T. (2012). *Etude stratégique du mix energétique pour la production d'electricité en Tunisie: rapport final.*
7. La Contribution Determinee au niveau National (CDN) de la Tunisie. https://www4.unfccc.int/sites/ndcstaging/PublishedDocuments/Tunisia%20First/Tunisia%20Update%20NDC-french.pdf
8. The United Nations Development Program (UNDP). (2022). https://www.undp.org/about-us
9. Ministère de l'Energie, des Mines et de la Transition Energétique. (2022). *Aperçu sur le secteur de l'énergie.* https://www.energiemines.gov.tn/fr/themes/energie/#:~:text=Au%20cours%20de%20la%20p%C3%A9riode,'%C3%A9quivalent%20p%C3%A9trole%20(tep)%20.
10. United Nations. (2022). *Digitization update—Conv on climate change.* https://www.un.org/en/library/digitization-update-conv-climate-change
11. United Nations. (2022). *The Paris agreement.* https://www.un.org/en/climatechange/paris-agreement
12. ANME. (2022). *Politiques d'atténuation dans le secteur de l'énergie.* http://www.anme.tn/?q=fr/content/politiques-dattenuation-dans-le-secteur-de-lenergie
13. Agence Nationale pour la Maitrise de l'Energie (ANME). http://www.anme.tn/
14. ANME. (2022). *Atlas Eolien.* http://www.anme.tn/?q=fr/projcts/eolien/atlas-eolien
15. Deutsche Gesellschaft für Internationale Zusammenarbeit (GIZ). https://www.giz.de/
16. GIZ. (2022). *Renforcement du marché solaire en Tunisie.* https://www.giz.de/en/worldwide/27358.html].
17. Plan solaire Tunisien (PST). (2015). https://www.energiemines.gov.tn/fileadmin/user_upload/energies_renouvelables/PST_2015.pdf
18. Ministère de l'Energie, des Mines et de la Transition Energétique. (2022). https://www.energiemines.gov.tn/fr/tc/actualites/
19. ANME. (2022). *Raccordé en Mt-HT.* http://www.anme.tn/?q=fr/projets/solaire-photovoltaique/raccorde-en-mt-ht
20. Société Tunisienne de l'Electricité et du Gaz (STEG). (2022). https://www.steg.com.tn/
21. IRENA–International Renewable Energy Agency. (2019). *Future of solar photovoltaic: Deployment, investment, technology, grid integration, and socio-economic aspects.*
22. International Energy Agency (IEA). (2022). https://www.iea.org/
23. IEA. (2014). Technology roadmap, solar photovoltaic. In *Energy* (2014 edition). https://iea.blob.core.windows.net/assets/e78cd964-9859-48c8-89b5-81fb4a1423b3/TechnologyRoadmapSolarPhotovoltaicEnergy_2014edition.pdf
24. Rauf, A., Al-Awami, A. T., Kassas, M., & Khalid, M. (2021). Optimal sizing and cost minimization of solar photovoltaic power system considering economical perspectives and net metering schemes. *Electronics, 10*(21), 2713.

25. GIZ. (2011). *Etude du développement des systèmes solaires thermiques collectifs dans le résidentiel.* https://energypedia.info/images/3/3e/FR_CES_Residence_Collectif_CAMI_2 011_GIZ.pdf

26. National Institute of Statistics. (2021). *Characteristics of housing.* http://www.ins.tn/statistiq ues/103

27. STEG. (2019). *Annual report 2019.* https://www.steg.com.tn/fr/institutionnel/publication/rap port_act2019/Rapport_Annuel_steg_2019_fr.pdf

28. NAMA Facility. (2022). https://www.nama-facility.org/.

29. Kimball, R., & Ross, M. (2011). *The data warehouse toolkit: The complete guide to dimensional modeling.* Wiley.

30. Inmon, W. H. (2005). *Building the data warehouse.* Wiley.

31. Kimball, R. (1996). *The data warehouse toolkit: Practical techniques for building dimensional data warehouses.* Wiley.

32. Wang, J., Yang, Y., Wang, T., Sherratt, R. S., & Zhang, J. (2020). Big data service architecture: A survey. *Journal of Internet Technology, 21*(2), 393–405.

33. Patel, M., & Patel, D. B. (2021). Progressive growth of ETL tools: A literature review of past to equip future. *Rising Threats in Expert Applications and Solutions,* 389–398.

34. Nayak, L. S., Das, K., Hota, S., & Sahu, B. J. R. (2022). Implementation of data warehouse: An improved data-driven decision-making approach. In *Intelligent and cloud computing* (pp. 419–427). Singapore.

35. Badgujar, A. D., Kadam, S. S., Zambare, M. M., & Kulkarni, S. R. (2022). A comparative study: Business intelligence tools. *International Journal of Research in Engineering, Science and Management, 5*(1), 116–118.

Clean Energy Management Based on Internet of Things and Sensor Networks for Climate Change Problems

Yasmine S. Moemen⊙, Heba Alshater⊙, and Ibrahim El-Tantawy El-Sayed

1 Introduction

Since the beginning of the industrial revolution [1], it has been abundantly clear that human activities have hastened the progression of climate change. Since the beginning of industrialization, there has been a roughly fifty percent rise in anthropogenic carbon dioxide in the atmosphere. The primary contributor has been the use of fossil fuels for the generation of energy, the operation of transportation networks, and the processing of industrial goods [2]. The residential building industry and the commercial construction industry are major consumers of energy. In 2018, the built environment was responsible for around 40% of the world's total energy consumption and 40% of the planet's greenhouse gas emissions. The annual emissions from the building sector reached a new high in 2018, having climbed by 2% from the previous year. This took place even though overall energy usage was increasing by 1%. Guidelines 14 and 15, the International Performance Measurement and Verification Protocol (IPMVP), the Federal Energy Management Program (FEMP), and

Y. S. Moemen (✉)
Clinical Pathology Department, National Liver Institute, Menoufia University, Menoufia, Egypt
e-mail: yasmine.moemen@gmail.com; yasmine_moemen@liver.menofia.edu.eg
URL: http://www.egyptscience.net

H. Alshater
Department of Forensic Medicine and Clinical Toxicology, University Hospital, Menoufia University, Shebin El-Kom, Egypt
e-mail: Heba.alshater@med.menofia.edu.eg
URL: http://www.egyptscience.net

I. E.-T. El-Sayed
Chemistry Department, Faculty of Science, Menoufia University, Menoufia, Egypt
URL: http://www.egyptscience.net

Y. S. Moemen · H. Alshater · I. E.-T. El-Sayed
Scientific Research Group in Egypt (SRGE), Giza, Egypt

© The Author(s), under exclusive license to Springer Nature Switzerland AG 2023
A. E. Hassanien and A. Darwish (eds.), *The Power of Data: Driving Climate Change with Data Science and Artificial Intelligence Innovations*, Studies in Big Data 118,
https://doi.org/10.1007/978-3-031-22456-0_7

the Uniform Methods Project of the Department of Energy (DOE) (UMP) [3]. They establish guidelines for determining how much energy is used, how much energy is demanded and how much water is used [4]. However, the programs do not target any metrics relevant to reducing the effects of climate change. It could be a game-changer if the building sector were to transform energy savings into pollution reductions. It is possible to do so by locating a generally accepted method to which all relevant parties agree and which can provide trustworthy auditable reporting and verification in 2017 and 7% higher than in 2010 [5]. These concerning figures are attributable, to a significant degree, to the expansion of both the floor space of worldwide buildings and the human population. Although there is the continued implementation of energy-efficiency methods in buildings, these strategies are insufficient to keep up with the demand for energy [6]. As a result, energy-efficiency methods for buildings and management strategies for those buildings are vital to the process of mitigating the effects of climate change. The building industry launched several auditing schemes to keep track of how much energy was being used. The American Society of Heating, Refrigeration, and Air-conditioning Engineers is one example of such an organization (ASHRAE).

The carbon credit market [7] relies heavily on the measuring, reporting and verification (MRV) system. An important application area for MRV technology is building energy performance (BEP) monitoring. The existing BEP MRV, on the other hand, cannot provide a reliable solution to the issues. The use of Blockchain technology can enhance this system's reliability. Blockchain is a distributed database that is instantly accessible to all participants in a network. Data recorded in a blockchain system are unchangeable, shareable, and traceable after they have been recorded. The BEP MRV system benefits from Blockchain's transparency, traceability, and affordability. As a result, the Emissions Trading System (ETS) design characteristics and MRV for carbon emissions [8] are great possibilities.

Human actions, particularly in the last half-century, are largely to blame for the rapid shifts in global climate [9, 10]. As long as heat-trapping gas emissions in the atmosphere and Earth's climate sensitivity remain high, the climate change expected to continue [11]. A rise of 1.3–1.9 F in average temperature has been recorded in the United States from 1895 to 2020 [12–14], with the greatest increase occurring after 1970. An 8-inch rise in global sea levels has been seen in this century since 1880, with a projected increase of up to 4 feet by the end of this century. Climate change has decreased the amount of ice covering the sea, land, and lakes. In recent years, the summer month on record has been broken on a regular making it impossible to effectively estimate climate change. As a result, the length of the growth seas growing risen and will continue to increase due to the interdependence of the growing season on the frost-free time [15–24]. There has also been an increase in the average amount of precipitation, as well as an increase in the intensity of the most extreme downpours and precipitation. The frequency of cold waves has decreased, but their strength has increased [25–31], as have other variations in extreme weather occurrence patterns. Carbon dioxide emissions from the atmosphere are being absorbed by oceans, and this is causing ocean acidification (a fall in ocean changes pH values) [32].

IoT is being hailed as a powerful weapon against climate change [33]. It can detect the amount of CO_2 and other greenhouse gases in the atmosphere and use that information to help uncover the underlying causes of climate change [34]. Real-time monitoring of greenhouse gas emissions from fossil fuel combustion is possible. So, monitoring carbon sequestration processes and rates can help to offset emissions by increasing the amount of carbon stored in forests. Climate IoT can be used to develop new atmospheric "things" and technology that can be used to permanently reduce CO_2 levels in the atmosphere.

Climate IoT can also be used to anticipate and prepare for climate change. Uncertainty in climate change can be reduced by using advanced sensors, communication networks, and models. Forecasts of climate change and ecosystem response can be made using IoT-enabled decision-making tools incorporating sensors' data. Increased greenhouse gas emissions may be empirically studied thanks to IoT technologies [35]. Additionally, it can simulate how an ecosystem will respond to various climates (both current and future). Environmental and atmospheric management procedures can now be better adapted, and new management approaches can be devised as a result of this new information being gleaned. Climate IoT can address short-term and long-term goals and application needs by exploiting current scientific and technical breakthroughs. This architecture supports improved knowledge and insight into global ecosystems. As we learn more about the Earth's system on a broader scale, we'll be better able to estimate water resources, forecast weather patterns, and assess the health of ecosystems. As a result of these reasons, we need to design applications that are useful to our community.

This chapter is organized as follows. Section 1 represents the ng introduction. In contrast, Sect. 2 contains greenhouse monitoring, Sect. 3 includes challenges related to energy use, Sect. 4: The IoT for sustainable energy, is also divided into a subsection, and Sect. 5 sensors for the transmission system, Sect. 6: Meters with Internet-enabled functions, Sect. 7: sensing of the solar and wind fields and conclusion.

2 Greenhouse Monitoring

Plants that need carefully controlled temperatures and humidity levels are typically grown in a greenhouse since it is an enclosed structure with walls and a roof composed primarily of transparent material, such as glass. These buildings can be as small as sheds or as large as factories, and their sizes range. A cold frame is like a tiny greenhouse in appearance. The temperature inside a greenhouse exposed to sunlight will become noticeably warmer than the temperature of the surrounding environment, which will shield the contents of the greenhouse from the cold. There are a lot of commercial glass greenhouses and hot houses out there, and many are high-tech production facilities for growing flowers and vegetables. The glass greenhouses are outfitted with various machineries, such as screening installations, heating and cooling systems, and lighting, which a computer may manage to provide the ideal environment for developing plants. After that, various methods are utilized to assess the optimality degrees and comfort ratio of the micro-climate within the greenhouse (i.e., air temperature, relative humidity, and vapor pressure deficit) before cultivating a particular crop. This is done to lower the production risk. The conversion of chemical energy into electrical energy is the primary function of an electrical device known as a battery. Batteries can be broken down into various subcategories determined by their intended function; these subcategories are then utilized in various electronic and electrical products. Electrical batteries include a variety of substances, including mercury and lead compounds, among others; lead is one of the most toxic elements found in nature, and batteries do not contribute to environmental health [36].

Several researchers developed a new kind of battery called the bio battery. This battery draws its power from various organic compounds, including carbohydrates, amino acids, and enzymes. The sugar-digesting enzymes and the mediator make up the anode, and the oxygen-reducing enzymes and the mediator make up the cathode. These batteries have the potential to function with as many types as possible of energy sources within the greenhouse monitoring system. In addition to these dangers, there is also the possibility that the battery could explode or that there will be a leak of chemicals. Bio-battery have been developed as a solution to this issue. This battery lessens the effect these chemicals have on the environment and thus confers a significant benefit to humans.

Since the 1990s, many monitoring systems for greenhouses and the environment have been developed. However, these monitoring systems have been left behind due to a lack of knowledge and cost and implementation constraints. The use of this technology has the potential to assist in the expansion of agriculture within a managed procedures like increasing crops yield, rationing water use and other resources, it is required for plants to have the necessary environmental conditions, such as respiration. Plants' ability to absorb water might be hindered when soil temperatures are low. The amount of sunshine a plant receives can influence its rate of development. Low relative humidity causes an increase in transpiration, leading to a water shortage in plants. By automating the data collection process regarding the soil conditions and the numerous environmental parameters that influence plant growth, it is possible to collect information with fewer requirements for human labor. Automatically controlling all the factors that affect plant growth is also difficult because it is expensive, and some physical factors are interrelated. For instance, temperature and humidity are related so that when temperature increases, humidity decreases; therefore, controlling both together is difficult because they are interrelated. It is possible to utilize a wireless sensor network to gather the data from one place to the next in a greenhouse, which is necessary since the temperature and humidity levels inside the greenhouse need to be constantly monitored [36].

The sensor will measure the data coming from the greenhouse and then transfer the data it has acquired to the receiver. The construction of greenhouse monitoring systems is getting further along as each day passes. When utilizing this system, the monitoring process is simplified, and additional savings are realized in the areas of installation cost and ongoing maintenance cost.

3 Challenges Related to Energy Use

One of the fastest-growing fields of study is energy management, which is a rewarding career choice. The economic and environmental benefits of proactive energy system assessment and management can be realized. An energy manager's job is to evaluate how much energy is being used and then make modifications to make the system more efficient. Regarding planning for efficiency, it's common for energy management to focus on machinery, equipment, buildings, and other physical structures and processes[37].

Energy managers are responsible for evaluating and enhancing the efficiency of a company's present plans and processes to reduce environmental impact while simultaneously increasing profits. These include hydropower, solar battery storage and energy conversion, electrical networks, and petroleum processing and utilization.

Energy management is becoming increasingly vital as the world's natural resources are depleted. Making use of valuable resources more effectively requires the adoption of efficient systems. Emissions are a problem throughout the entire energy supply chain, including extraction, conversion, transportation, and distribution. Increasing a system's energy efficiency makes sense because it lowers costs and maximizes the inherent value of the resources being used.

Technology that considers environmental impact first and seeks to lessen it is a rapidly expanding industry. Changemakers are required in this industry to produce better energy management systems and invent new ways of processing, extracting, transporting, and so on.

We must think beyond the box for both your career advancement and that of our planet. We aim to show that climate protection and energy efficiency can also be commercially beneficial, and we hope to increase awareness in other companies. On top of that, we have close ties to the political sector and can give our expertise and experience to the legislators." Our current climate change dilemma can only be solved if we improve our energy management systems [37].

There are new emerging difficulties that future energy systems should handle by making use of the modern breakthroughs in energy technology [38, 39]. This is because these advances are developing simultaneously as the modern advanced energy technologies. The following discussion will focus on the past, present, and potential future difficulties that the community faces regarding energy. These difficulties affect the power-producing system's capacity and cause interruptions in the energy distribution system.

The current energy system relies substantially on water to function. However, due to the inconsistent supply of water (both in the short term and over longer periods), innovative energy production methods are required. There are no safety precautions in place for the high-voltage transmission lines. These likewise operate excessively and are not being used for their intended purpose. In addition, transmission loss is a significant problem that can result in power interruptions and blackouts. Because there is greater water availability in coastal areas, many energy plants are situated there.

Nevertheless, coastal infrastructure and energy facilities and infrastructure are being impacted by rising sea levels and high tides, heavy downpours, and flooding caused by storm surges [40–44]. Loss of productivity in urban and industrial locations, where the power outage lasted for an extended period, is related to a decline in the number of businesses and the overall economy. High energy demands and a commensurate increase in electricity usage are caused by the intense heat waves and temperatures prevailing this summer. It is anticipated that the energy demand will rise as a result of peak loads [45].

Increasing greenhouse gas emissions is one of the challenges of increased energy usage [46]. Carbon dioxide (CO_2), methane (CH_4), and nitrous oxide (N_2O) are some of the gases that are released as a result of these emissions [47]. After the transportation sector, responsible for the vast majority of emissions, these are the second highest. The energy consumed by information technology, mobile devices, and computers is expected to rise [48]. The production of electric vehicles (EV) is leading to an increase in the demand for energy, which is necessary to meet the requirements of EVs [49]. A mismatch between the demand and capacities of energy systems and the complicated energy needs of businesses and communities is another difficulty tied to energy infrastructure [38].

4 The Internet of Things for Sustainable Energy

It is abundantly clear from the discussion of energy and sustainability that universal access to energy cannot be achieved without the implementation of appropriate technological measures. The application of technology can make it possible to design resilient solutions for reliable, low-cost energy access, which can improve the performance and operation of the energy systems that are now in place. Because of this, the requirement for the community to have access to low-cost energy can be satisfied by utilizing the sensing and communication technologies of the next generation [50]. To satisfy this fundamental requirement of human existence, IoT must be developed into a system that can effectively deliver economic and efficient services.

Regarding sustainable energy systems, the IoT is envisioned as a way to connect all of the electrical grid's energy objects, service supply chains, and human capital using cutting-edge technology to meet the century's future needs and access challenges to clean energy sources. This paradigm is essential to connect a wide range of energy technologies and new solutions on a global scale. IoT for sustainable energy has the potential to make the current energy infrastructure more sustainable and resilient. Energy infrastructure and technologies that are safe, inventive, and efficient are among the capabilities that it possesses. By easing the implementation of large-scale renewable and clean energy solutions, IoT offers a variety of ways to provide low-cost energy sources to people worldwide [50].

Sustainability IoT is all about smart grids, a 21st-century technological marvel. Combining IoT autonomy with efficient grid management can increase production and consumption in the long term. Solar and wind power efficiency can be improved via real-time monitoring of renewable energy supplies and environmental monitoring. To increase supply, these can be connected to the grid. To reduce the use of fossil fuels, distributed and low-loss smart microgrids will be implemented, and included in sustainability are.

1. Generation wind, solar, natural gas, water, renewables, and coal.
2. Phasor measurement unit, transmission, and phasor measurement. Data collection and monitoring under supervision (SCADA).
3. Control of voltage, distribution, and smart and microgrid systems.
4. Work order and invoice management are also included in this category.
5. Providers of goods and services to the general public include the management of loads, bulks, and outages.

4.1 Coal-Plant Sensors

To meet the ever-growing demand for ecologically friendly, reliable, and adaptive power generation. Coal power plant control systems have experienced continual improvement. It has become necessary to use online monitoring technologies and more advanced algorithms to optimize the combustion process to manage multivariable systems. Coal and airflow sensors, along with imaging and spectral analysis of the flame, can help improve stoichiometric management. It is also possible to map the furnace's hot zones using in-situ laser absorption spectroscopy. One of the current plant control strategies that modern plant control systems can use is artificial intelligence, which mimics the behaviours of expert operators and uses complicated empirical models created from operational data to identify the optimum control response. These advanced plant control systems can use a wide range of computational methods. New sensor technologies are being developed to improve control further and ensure that these sensors can withstand the harsh conditions of advanced coal plants and gasifiers.

Since optical fibre sensors may produce highly sensitive, distributed, and low noise measurements even when subjected to high temperatures, an increase in the use of optical technology is of particular importance. Microelectronic fabrication techniques and newly discovered high-temperature materials are currently being used to build miniature devices that provide a reliable and cost-effective solution for in-situ gas and parameter monitoring. The development process incorporates the use of several techniques and materials. Wireless communication and self-powering systems can make it easier to install distributed sensor networks and monitor inaccessible places with the help of these newly created sensors. In the future, self-organizing networks may play an increasingly essential role in future control systems [51].

Coal-fired power stations are essential to IoT-based systems for long-term sustainability as a source of fluctuating power. Monitoring in these facilities enhances combustion efficiency and permits self-optimization through sensors. It is possible to improve the performance of coal-fired power plants by employing modern stoi-chiometric control systems. various materials, including coal, flame, carbon, oxygen and flow sensors have been used in furnaces [52].

4.2 Oxygen Sensing

The amount of oxygen remaining after burning is a critical variable in combustion management, and oxygen sensing is essential for combustion monitoring in fossil-fuel-fired power plants [53]. Combustion air intake and distribution adjust this oxygen signal to an oxygen set point. On the other hand, changes in the rate of firing or other disturbances may need a change. Lowering the oxygen set point while avoiding incomplete combustion can improve combustion efficiency and reduce NOx emis-sions. The residual oxygen controls the process of burning. The rate of fire and the amount of air being drawn into the chamber can be altered—incomplete combustion and oxygen set-point optimization [54].

The voltage generated by platinum electrodes covered in catalytic platinum in the electrolyte is directly proportional to the gradient in oxygen concentration across the cell [51]. Ionic conduction can only take place at temperatures above 300 °C. Hence a special electric heater is needed to keep the zirconia between 700 and 750 °C.

Oxygen sensors can be found in various forms, but the most common is an elec-trochemical zirconia-based sensor. Zinc oxide is used as a solid electrolyte between the sample gas and the air as a reference for this device's oxidation detection [55].

Using probes, it is possible to install Zirconia sensors directly into the flue gas. Ceramic or stainless-steel casings protect these sensors from high temperatures and fly ash diffused through a filter. The use of temperature-resistant ceramic shielding can withstand temperatures as high as 1400 °C, making them suitable for the furnace's higher temperatures [51]. Due to the lack of a heater, these sensors from Rose-mount can operate at extremely high temperatures. A hermetically sealed metallic reference is used as a substitute for air to ensure that measurements are free of drift. Zirconia sensors have many benefits, including that their inverse logarithmic response increases accuracy as oxygen concentration decreases. As a result, zirconia sensors are ideal for use in environments with low amounts of oxygen following burning [51]. Only one or a few sensors are used in most coal boilers. Tempera-tures of 300–400 °C are common between the economizer and the air preheater, where these sensors are frequently installed. Because of its proximity to the furnace and distance from the probe, this location necessitates materials that can tolerate high temperatures to regulate combustion. This can make it difficult to distinguish between the flue gases produced by distinct burners due to air incursion in convective passes. When attempting to optimize the furnace's oxygen set point, it's typical to run across problems like an inaccurate picture of the furnace's actual oxygen levels due to improper monitoring. In addition to paramagnetic analyzers, which make use of oxygen being pulled to a magnetic field, extractive oxygen analyzers can also use zirconia sensors [51]. Oxygen movement can be detected in several ways, such as with flow sensors or by the torque exerted on a revolving pair of nitrogen-filled glass spheres after exposure to an intense magnetic field [51]. Servomex and Yokogawa developed these methods. Unlike zirconia sensors, this measurement method is unaffected by combustible gases, which are known to artificially reduce the signal by interacting with oxygen. One of the advantages of this form of measurement is that it provides a more accurate picture of the situation. An additional layer of complexity and slower response times are incurred by installing a gas sampling system.

Air preheater and economizer were employed in zirconia-based electrochemical sensors. These sensors use platinum electrodes capable of separating and absorbing oxygen into electrons and ions [56].

Paramagnetic sensors can measure oxygen because of the strong magnetic field. It uses two nitrogen-filled glasses to cause suspension rotation, which photocells sense. It's less sensitive to combustion gases [57].

4.3 Carbon Monoxide Sensors

The carbon monoxide (CO) concentration in the flue gases can serve as an extremely helpful control variable within the furnace. It should ideally be maintained at a level lower than 200 parts per million (ppm). CO is the most sensitive and accurate indicator of incomplete combustion [58, 59]. Suppose an unwanted rise in CO levels, sometimes known as a CO "breakthrough," is detected. In that case, the excess oxygen set point can be lowered to a more appropriate level, and the extra air can be adjusted accordingly. Alternatively, a CO sensor that is more sensitive could be used as a control variable for the furnace itself, particularly in regards to optimizing the oxygen set point.

In most cases, CO detection in coal furnaces relies on either infrared absorption or electronic sensors that rely on catalytic combustion as their primary method of operation. This latter group uses a conductive element covered with a catalyst that encourages combustion, such as platinum. The conductive element is heated, which raises its resistance, as CO and other combustibles are oxidized on the catalyst. In coal boilers, the other combustibles are often negligible compared to the CO [60]. The most prevalent use of this technique is found in catalytic bead sensors, which involve coating a conductive filament with a bead of catalyst. These sensors can be found in GE, ABB, and Emerson/Rosemount devices. These devices are too sensitive to be used in situ and require sample extraction; nonetheless, they are capable of being "close linked," in which sample conditioning consists merely of the filtration of particulate matter [61, 62]. Although the Rosemount sensor promises to be resistant to sulfur, the sensitivity of catalytic bead sensors in applications involving coal plants to catalyst poisoning by SO_2 is one of the sensors' weaknesses. Servomex produces a thick film thermistor, which is an alternate application of the idea of catalytic combustion. This type of thermistor consists of thin conductive tracks formed on a ceramic substrate and coated with a layer of CO-sensitive catalyst. This design is also applicable in a close-coupled arrangement, and it boasts a high degree of precision as well as resistance to the poisoning of catalysts [63].

The IR analysis of the flue gas CO content can either be extractive, in which case the flue gas is removed from the furnace and placed in a sample cell for analysis, or it can be in situ, in which case an IR source and detector are placed on either side of the flue gas duct. The entire flue gas volume acts as a sample cell. In-situ devices Rosemount and SICK [64, 65] make use a technique called gas filter correlation. During this technique, a portion of the detected beam is passed through a vessel filled with pure CO. This saturates the CO absorption signal and establishes a baseline for the interference caused by the absorption of other species [64, 65]. Even though they offer a usable average over a full portion of the furnace, line-of-sight measurements are susceptible to high amounts of particles and temperatures much higher than 600 °C. There is no feasible method of calibrating the measurement, and thermal expansion and vibration are potential factors that could throw off the alignment of the source and receiver, necessitating signal filtering. A dual-pass arrangement, in which a furnace probe is utilized to reflect the beam to a combined source and detector unit, is one method that can be utilized to alleviate alignment concerns. The use of tuneable diode lasers as the source is a relatively recent development in line-of-sight infrared technology. This allows for greater accuracy and monitoring in areas with high temperatures.

4.4 Flame Detection

The safety of pulverized coal combustion depends on flame sensing in coal-fired power plants. These sensors measure the flames' infrared, visible, and ultraviolet light frequencies. These flame stoichiometry and temperature data increase combustion [66].

Optical flame detectors have been placed on each burner to ensure that pulverized coal is properly burned safely. The amplitude and frequency (also called flicker) of selected visible, infrared, or ultraviolet frequencies generated by the flame are commonly measured by these instruments. This information can be used to improve the combustion process by analysing it further [67–69].

For example, ABB's Advisor series of flame scanners provides additional information on the quality of the flame in addition to the conventional requirement of just monitoring its presence. Burners that aren't working properly can be identified with Flame Doctor®, a portable diagnostic device that uses signals from existing flame scanners to identify them. It is possible to detect different abnormalities in flame quality and optimize the air–fuel ratio based on these deviations using software that recognizes patterns and mathematics developed from the chaos theory [70].

In addition, video cameras can capture photographs of the flame in real time. With the right processing software, even a higher quantity of information regarding the quality and consistency of the flame can be derived from these images.

4.5 Sensing Coal Flow

Conventionally, the gravimetric federate of coal to the pulveriser mills is used as the sole metric for monitoring coal flow. This federates directly controlled by the boiler fire rate and the amount of load that the plant is required to produce.

It is only possible to check the distribution of coal across burners on an ad hoc basis using sample probe readings, which are not always accurate and are not carried out concurrently on different pipelines [71]. Even though the actual coal flow rate is nearly always lower, it is normal practice to draw coal from the coal pipe at the same rate as the airflow.

Online flow sensors on coal pipelines have expanded in use due to the need for better control of individual burner stoichiometry. There are currently a variety of technologies that can be used for this purpose. Coal charge is detected electrostatically by electrodes and correlated with the same data at a downstream sensor to derive the time-of-flight between two points and, as a result, a computation of coal velocity by using ABB and Greenbank's PFMaster system [51, 72].

It is used to balance the distribution of coal over a group of burners, together with information produced from the overall charge detected. The PFMaster can detect pulsed flow behavior due to its quick response time. Because the electrodes are designed to be flush with the pipe, erosion will not occur because of this feature. The PF-Flo uses electrostatic cross-correlation and the Me control coal, manufactured by Air Monitor Corp. and Protection, to determine coal velocity. To get an accurate mass flow measurement, they use a microwave resonance approach that considers coal density in the pipe [73]. Burner pipes at the Stigsnaes Power Plant in Denmark were fitted with Me control Coal sensors, which resulted in a 30% reduction in oxygen set point, 44% reduction in NOx, and an efficiency improvement of 1.3%.

Others, like EUtech's EUcoalflow and MIC's Coal Flow Analyzer, use microwave signals with an increased frequency to measure the amount of coal flowing through the system. Two or three non-intrusive microwave transceivers installed around the circumference of a pipe can be used to transform the time-dependent intensity of microwaves reflected by moving coal particles into an absolute mass flow rate. In this way, the flow rate may be determined. With these sensors, EUtech's whole air–fuel ratio optimization technique is expected to yield efficiency benefits of 0.3–1%, as well as reductions in emissions and lagged performance, according to the company [51].

Since they are non-intrusive and less affected by ambient factors, including temperature, humidity, and other charge sources, optical image-based techniques have recently challenged older methods. Due to the increasing accessibility of digital imaging equipment, this is already a reality [74]. Coal particles are lit by high-intensity LEDs using CCD digital video cameras in these devices. They can determine the particle concentration and the particle velocity by analysing the blurriness of the photographs. Deposits can build over time, making it difficult to see through the coal pipe's transparent window.

4.6 Sensing Airflow

When it comes to controlling the combustion process, one of the most important parameters to adjust is the combustion air flow rate into the furnace. Additionally, the flow of primary air into the pulverizer mills needs to be maintained within a specific range that keeps the coal in suspension while minimizing erosion and the production of NOx. Venturi flow meters, which measure the decrease in fluid pressure that occurs as air travels through a confined piece of pipe, or Pitot tubes, which measure the pressure that air builds up when it is allowed to come to a complete stop, are generally used to monitor both of these air movements ("Air Monitor Corp, Flow Measuring and Control Stations" 2015). Air Monitor Corporation's IBAM system, which uses pitot tubes in each burner's combustion chamber, is proof that this technology can be used on even the smallest of burners. Fechheimer Pitot tubes, or flow straightening devices, may be necessary to account for the non-axial flow components of turbulent air. Short or curved duct portions are common places to find these parts (Air Monitor Corp, Rosemount). Both Pitot tubes and Venturis must have self-purging mechanisms in dirty air to function efficiently. If the principal air flow is being evaluated, these changes are especially critical because the ductwork is often short and the flow is contaminated by fly ash from regenerative heaters. The velocity of entrained particulates can also be used as a proxy for the airflow, and the charge signals created by these particles can be used to get the velocity of the entrained particulates in the same manner as for coal flow [75].

Volume flow meters include devices such as Pitot tubes and Venturis. This flow meter necessitates additional temperature and pressure measurements to accurately establish the density of the air and, therefore, the total mass flow. As a result, temperature inhomogeneities, such as those caused by the attemperator of primary air, might introduce errors. An alternative to mass flow meters is thermal mass flow meters. To measure the mass flow, these meters use the convective cooling effect of flowing air on a hot object [76, 77].

Additionally, compared to pitot tubes, these are more accurate at low flows. Venturis, on the other hand, cause the duct to experience energy-intensive pressure drops; this design avoids this problem. Fouling was a common concern with early thermal anemometers, but newer designs, such as Kurz Instruments' thermal mass insertion meters that operate at far higher temperatures than the air around them, eliminate this issue.

Optical flow meters, a new type of coal flow meter, are non-invasive and less sensitive to external factors, making them ideal for monitoring coal flow ("Optical Scientific, Optical Flow Sensors." 2015). Optical scintillation happens when light is diffracted due to localized fluctuations in air temperature and density and they can take advantage of this.

It is also possible to estimate burner airflow by combining existing flow measurements with a physical system airflow model. This is one method for determining burner airflow. A "soft sensor" utilized by EUtech to monitor air flow at all points in the hydraulic network of the plant and adapt to changing inputs, such as damper position, employs this strategy. Everything can be done using EUtech's 'EUsoft air' A robust PLC generates data from the soft sensors in real-time and sends it directly to the DCS [78].

4.7 Ash Carbon Sensing

Carbon ash concentration indicates combustion efficiency [79] Less than 20% is kept. Microwaves are used to assess carbon concentration. Depending on the dielectric constant, carbon's high permittivity absorbs EM radiation. Resonant cavity sensors detect frequency fluctuations.

4.8 Temperature-Sensing Gases

In most cases, the results of gas sensing [80] performed in a particular area do not accurately represent the gas concentration. For this reason, arrays of gas sensors, both linear and planar, are utilized to obtain a comprehensive picture. Another innovation for sensing the quantities of flue gases with a very high level of accuracy is the tunable diode laser absorption spectroscopy. Monitoring the temperature of the furnace's exit gas is also very significant for controlling the furnace. The sensing of temperature can be accomplished by utilizing several strategies that are covered in the section on nuclear reactors. In addition, nitrogen oxide monitoring is carried out in the plant to detect the presence of nitrogen oxides.

5 Sensors for the Transmission System

Sensing grid transmission systems is critical for various reasons [81]. The sensing technologies are either fully developed or in the process of being developed. The following applications are discussed.

5.1 Methods for Sensing at Substations

(1) **Monitoring Potential Discharge at Substations**: To prevent catastrophic fail- ures, it is essential to monitor any potential discharge that may occur at substa- tions [82]. Antenna arrays are being utilized to assess, locate, and identify components that are contributing to discharge at present. In addition to that, the approaches of 3D acoustic emissions are currently being utilized for discharge

sensing in transformers. In addition to this, 3D acoustics can be used to detect bubbling sources as well as gas sources.

(2) **Video Imaging**: In this method, IR tomographic cameras are utilized to make thermal video images of substation components. This approach is known as "Video Imaging."

(3) **Metal Insulated Semiconducting (MIS) Gas in Oil Sensor**: this detects the presence of gas in the oil. In this method, a hydrogen sensor that is not very expensive is used to monitor the concentrations of H_2 and C_2H_2 in the headspace and oil of transformers. The MIS gas sensor is built on a chip during manufacturing [83]. This sense is also utilized to determine whether or not cable oil contains hydrogen and possible acetylene.

(4) **Sensing-based on fiber optic technology.** Two varieties of sense are based on fibre optics: acoustic and gas. Fibre optics cable is used to check for discharge in the stress zones of the transformers in sound. The presence of a gas at the end of the fibre optics is analysed in the second method [84] which is utilized to detect early stages of degradation and failure in high-risk locations.

(5) **Frequency Domain Analysis**: The device's functionality is based on the frequency domain analysis of the transformers as its underlying concept. In the FDR method, the configuration modifications of the transformers are identified by analysing the fluctuations in frequency response data. These measurements are obtained in a shady manner by using spontaneous transients [85].

(6) **Sensing Gas in System Load Tap Changer (LTC)**: This sensing approach can measure the LTC gas ratios without requiring individual gas measurements [86]. It can do so because it can monitor the LTC gas ratios.

(7) **Radio Frequency**: RF-based sensing technologies are utilized to detect leakages of the current levels to provide information regarding insulation washing and flash-over for a wide variety of insulation types [87]. In addition, they can perform wireless or remote identification of high-risk components (for example, acoustics-based internal discharge, current, jaw temperature of disconnect, and density of sulfur hexafluoride). For the goal of this endeavour, both the timing and magnitudes of fault currents traveling through the shield wires are utilized.

5.2 Sensing of Overhead Lines

The following is a discussion of the various techniques for sensing overhead lines.

1. The current sensing and temperature sensing methodologies are applied in overhead transmission to sense the temperature of connectors, the current magnitudes, and the compression of conductors such as dead ends and splices. As a result, a histogram is constructed to evaluate the loss and locate components subject to significant stress. These sensors can harvest energy from the considerable magnetic field in the vicinity of the line [88].

2. In an environment analogous to a substation, RF methods are utilized to assess the leakage of currents connected with overhead insulators. When pinpointing

the precise site of faults, it is necessary to measure not only the amount of current but also the amount of time it has been flowing through the shield wire. Regarding the illumination current distributions, the same measurements are taken [89].

3. The surge sensor is employed to measure and log surges and the overall charge sensed [90].
4. The transmission structure's sensing is accomplished by utilizing the sensing of the environment data and the image processing for decision support systems. As a result, various situations, such as an unknown outage and actions by birds of prey, can be identified in real-time [91].

6 Meters with Internet-Enabled Functions

In the IoT for renewable energy, smart meters can also be considered a form of sensor. The advanced metering infrastructure is built on top of smart meters, which are the primary constituents of this infrastructure (AMI).

These utilize a variety of communication channels to establish a connection between clients and service providers [92]. Monitoring power flow in both directions uses smart meters in another capacity. Consequently, using smart meters makes it possible to implement dynamic invoicing, load monitoring, and remote capabilities.

7 Sensing of the Solar and Wind Fields

Real-time sensing of these environmental factors is necessary for the effective operation of the energy generating process [93], as it is required for the reliable integration of solar [94, 95] and wind energy [96–98] resources in the IoT for sustainable energy. Solar irradiation and wind speed are both measured during these environmental sensing procedures. Sensing weather-related factors of this type in sustainable energy systems has the enormous potential to deliver a greater variety of energy sources to power systems.

8 Conclusion

Climate change is closely related to energy management. A lot of challenges were related to energy use. Still, The IoT offered solutions for sustainable energy, like the generation of wind, solar, and renewables which were performed through IoT sensing like coal-plant sensors, oxygen sensing, Co sensing, and multiple sensors or detectors. Meters with Internet-enabled functions are considered sensors. Aside from wind speed and its energy generation, it can be used as a monitor for solar irradiation. In sustainable energy systems, weather-related parameters of this type can deliver a wider range of energy sources to power systems.

References

1. Wrigley, E. A. (2018). Reconsidering the industrial revolution: England and Wales. *The Journal of Interdisciplinary History, 49*(1), 9–42.
2. Manabe, S., & Broccoli, A. J. (2020). *Beyond global warming: How numerical models revealed the secrets of climate change.* Princeton University Press.
3. Li, M., Haeri, H., & Reynolds, A. (2017). Introduction. In: *The uniform methods project: Methods for determining energy-efficiency savings for specific measures.* National Renewable Energy Lab (NREL).
4. Woo, J., Fatima, R., Kibert, C. J., Newman, R. E., Tian, Y., & Srinivasan, R. S. (2021). Applying blockchain technology for building energy performance measurement, reporting, and verification (MRV) and the carbon credit market: A review of the literature. *Building and Environment, 205*, 108199.
5. Zhang, F., Zhang, W., Li, M., Zhang, Y., Li, F., & Li, C. (2017). Is crop biomass and soil carbon storage sustainable with long-term application of full plastic film mulching under future climate change? *Agricultural Systems, 150*, 67–77.
6. GlobalABC. (2019). *Global alliance for buildings and construction, International Energy Agency and United Nations Environment Programme.* 2019 Global Status Report for Buildings and Construction: Towards a Zero-Emissions, Efficient and Resilient Buildings and Construction Sector.
7. Singh, N., Finnegan, J., Levin, K., Rich, D., Sotos, M., Tirpak, D., & Wood, D. (2016). *MRV 101: Understanding measurement, reporting, and verification of climate change mitigation.*
8. Braden, S. (2019). Blockchain potentials and limitations for selected climate policy instruments. *Retrieved 5*(9), 2020
9. Karl, T. R, Melillo, J. M., Peterson, T. C., & Hassol, S. J. (2009). *Global climate change impacts in the United States.* Cambridge University Press.
10. Sands, R. D., & Edmonds, J. A. (2005). Climate change impacts for the conterminous USA: An integrated assessment. In *Climate change impacts for the conterminous USA* (pp. 127–50). Springer.
11. Nakicenovic, N., Alcamo, J., Davis, G., de Vries, B., Fenhann, J., Gaffin, S., Gregory, K., Grubler, A., Jung, T. Y., & Kram, T. (2000) Special report on emissions scenarios.
12. Alexander, L. V., Zhang, X., Peterson, T. C., Caesar, J., Gleason, B., Tank, A. M. G. K.. Haylock, M., Collins, D., Trewin, B., & Rahimzadeh, F. (2006) Global observed changes in daily climate extremes of temperature and precipitation. *Journal of Geophysical Research: Atmospheres, 111*(D5).
13. Rosenzweig, C., Rind, D., Lacis, A., & Peters, D. (2018). *Our warming planet: Topics in climate dynamics* (Vol. 1). World Scientific.
14. Council, National Research. (2011). *Climate stabilization targets: Emissions, concentrations, and impacts over decades to millennia.* National Academies Press.
15. Adams, R. M., Hurd, B. H., Lenhart, S., & Leary, N. (1998). Effects of global climate change on agriculture: An interpretative review. *Climate Research, 11*(1), 19–30.
16. Darwin, R. (1995). *World agriculture and climate change: Economic adaptations.* US Department of Agriculture, Economic Research Service.
17. Easterling, W. E. (2011). *Guidelines for adapting agriculture to climate change.* Imperial College Press.
18. Hatfield, J. L., Boote, K. J., Kimball, B. A., Ziska, L. H., Izaurralde, R. C., Ort, D., Thomson, A. M., & Wolfe, D. (2011). Climate impacts on agriculture: Implications for crop production. *Agronomy Journal, 103*(2), 351–370.

19. Högström, U., & Smedman, A.-S. (2004). Accuracy of sonic anemometers: Laminar wind-tunnel calibrations compared to atmospheric in situ calibrations against a reference instrument. *Boundary-Layer Meteorology, 111*(1), 33–54.
20. Reilly, J., Tubiello, F., Bruce, D., Abler, R. D., Fuglie, K., Hollinger, S., Izaurralde, C., Jagtap, S., & Jones, J. (2003). US agriculture and climate change: New results. *Climatic Change, 57*(1), 43–67.
21. Smit, B., & Skinner, M. W. (2002). Adaptation options in agriculture to climate change: A typology. *Mitigation and Adaptation Strategies for Global Change, 7*(1), 85–114.
22. Wall, E., & Smit, B. (2005). Climate change adaptation in light of sustainable agriculture. *Journal of Sustainable Agriculture, 27*(1), 113–123.
23. Walthall, C. L., Hatfield, J., Backlund, P., Lengnick, L., Marshall, E., Walsh, M., Adkins, S., Aillery, M., Ainsworth, E. A., & Ammann, C. (2013). *Climate change and agriculture in the United States: Effects and adaptation.* United States Department of Agriculture, Agricultural Research Services.
24. Ziska, L. H. (2011). Climate change, carbon dioxide and global crop production: Food security and uncertainty. In: *Handbook on climate change and agriculture*, 9–31.
25. Alexander, M. A, Scott, J. D., Friedland, K. D., Mills, K. E., Nye, J. A., Pershing, A. J., & Thomas, A. C. (2018). Projected sea surface temperatures over the 21st century: Changes in the mean, variability and extremes for large marine ecosystem regions of Northern Oceans. *Elementa: Science of the Anthropocene, 6.*
26. Diez, J. M., D'Antonio, C. M., Dukes, J. S., Grosholz, E. D., Olden, J. D., Sorte, C. J. B., Blumenthal, D. M., Bradley, B. A., Early, R., & Ibáñez, I. (2012). Will extreme climatic events facilitate biological invasions? *Frontiers in Ecology and the Environment, 10*(5), 249–257.
27. Fowler, D. R., Mitchell, C. S., Brown, A., Pollock, T., Bratka, L. A., Paulson, J., Noller, A. C., Mauskapf, R., Oscanyan, K., & Vaidyanathan, A. (2013). Heat-related deaths after an extreme heat event—Four states, 2012, and United States, 1999–2009. *Morbidity and Mortality Weekly Report, 62*(22), 433.
28. Greene, S., Kalkstein, L. S., Mills, D. M., & Samenow, J. (2011). An examination of climate change on extreme heat events and climate-mortality relationships in large US cities. *Weather, Climate, and Society, 3*(4), 281–292.
29. Kunkel, K E. (2008). Observed changes in weather and climate extremes. In: *Weather and climate extremes in a changing climate: Regions of focus: North America, Hawaii, Caribbean, and US Pacific Islands* (pp. 35–80).
30. Peterson, T. C., Stott, P. A., & Herring, S. (2012). Explaining extreme events of 2011 from a climate perspective. *Bulletin of the American Meteorological Society, 93*(7), 1041–1067.
31. Stone, B., Hess, J. J., & Frumkin, H. (2010). Urban form and extreme heat events: Are sprawling cities more vulnerable to climate change than compact cities? *Environmental Health Perspectives, 118*(10), 1425–1428.
32. Doney, S. C., Fabry, V. J., Feely, R. A., & Kleypas, J. A. (2016). Ocean acidification: The other CO_2 problem. *Washington Journal of Environmental Law & Policy, 6*, 213.
33. Khan, R., Khan, S. U., Zaheer, R., & Khan, S. (2012). *Future internet: The Internet of Things architecture, possible applications and key challenges.* 2012 10th International Conference on Frontiers of Information Technology, pp. 257–260. IEEE.
34. Tao, F., Zuo, Y., Da Li, X., Lv, L., & Zhang, L. (2014). Internet of Things and BOM-based life cycle assessment of energy-saving and emission-reduction of products. *IEEE Transactions on Industrial Informatics, 10*(2), 1252–1261.
35. Privette, J. L., Fowler, C., Wick, G. A., Baldwin, D., & Emery, W. J. (1995). Effects of orbital drift on advanced very high resolution radiometer products: Normalized difference vegetation index and sea surface temperature. *Remote Sensing of Environment, 53*(3), 164–171.
36. Saba, S., Shetty, S. R., Kulsum, U., & Vinutha, M. (2020). IoT based green house automation system with power conservation using biofuel cell. *International Journal of Engineering Research & Technology, 8*(13).
37. Audsley, S. M. (2019). *Why study energy management?.* https://www.masterstudies.com/article/why-study-energy-management/

38. Gui, E. M., & MacGill, I. (2018). Typology of future clean energy communities: An exploratory structure, opportunities, and challenges. *Energy Research & Social Science, 35*, 94–107.
39. Tirado, M. C., Cohen, M. J., Aberman, N., Meerman, J., & Thompson, B. (2010). Addressing the challenges of climate change and biofuel production for food and nutrition security. *Food Research International, 43*(7), 1729–1744.
40. Cutter, S. L., Solecki, W., Bragado, N., Carmin, J., Fragkias, M., Ruth, M., & Wilbanks, T. J. (2014). Ch. 11: Urban systems, infrastructure, and vulnerability. *Climate Change Impacts in the United States: The Third National Climate Assessment, 10*(7930), J0F769GR.
41. Kessler, R. (2011). *Stormwater strategies: Cities prepare aging infrastructure for climate change*. National Institute of Environmental Health Sciences.
42. Means, I. I. I., Edward, G., Laugier, M. C., Daw, J. A., & Owen, D. M. (2010). Impacts of climate change on infrastructure planning and design: Past practices and future needs. *Journal-American Water Works Association, 102*(6), 56–65.
43. Sathaye, J., Dale, L., Larsen, P., Fitts, G., Koy, K., Lewis, S., & Lucena, A. (2012). *Estimating risk to California energy infrastructure from projected climate change*.
44. Wilbanks, T. J., & Fernandez, S. (2014). *Climate change and infrastructure, urban systems, and vulnerabilities: Technical report for the US Department of energy in support of the national climate assessment*. Island Press.
45. Wilhelmi, O. V., & Hayden, M. H. (2010). Connecting people and place: A new framework for reducing urban vulnerability to extreme heat. *Environmental Research Letters, 5*(1), 14021.
46. Williams, J. H., DeBenedictis, A., Ghanadan, R., Mahone, A., Moore, J., Morrow III, W. R., Price, S., & Torn, M. S. (2012). The technology path to deep greenhouse gas emissions cuts by 2050: The pivotal role of electricity. *Science, 335*(6064), 53–59.
47. EPA. (2011). Inventory of US greenhouse gas emissions and sinks: 1990–2009. In: *US environmental protection agency*.
48. Satyanarayanan, M., Gao, W., & Lucia, B. (2019). *The computing landscape of the 21st century*. Proceedings of the 20th International Workshop on Mobile Computing Systems and Applications, pp. 45–50.
49. Luin, B., Petelin, S., & Al-Mansour, F. (2019). Microsimulation of electric vehicle energy consumption. *Energy, 174*, 24–32.
50. Salam, A. (2020). Internet of Things for sustainable Community development: Introduction and overview. In *Internet of Things for sustainable community development* (pp. 1–31). Springer.
51. Lockwood, T. (2015). *Advanced sensors and smart controls for coal-fired power plant. CCC/251*, IEA Clean Coal Centre.
52. Teichert, H., Fernholz, T., & Ebert, V. (2003). Simultaneous in situ measurement of CO, H_2O, and gas temperatures in a full-sized coal-fired power plant by near-infrared diode lasers. *Applied Optics, 42*(12), 2043–2051.
53. Tan, Y., Croiset, E., Douglas, M. A., & Thambimuthu, K. V. (2006). Combustion characteristics of coal in a mixture of oxygen and recycled flue gas. *Fuel, 85*(4), 507–512.
54. Guth, U., & Wiemhöfer, H.-D. (2019). Gas sensors based on oxygen ion conducting metal oxides. In *Gas sensors based on conducting metal oxides* (pp. 13–60). Elsevier.
55. Wang, L., Zhang, Y., Zhou, X., & Zhang, Z. (2019). Sensitive dual sensing system for oxygen and pressure based on deep ultraviolet absorption spectroscopy. *Sensors and Actuators B: Chemical, 281*, 514–519.
56. Rocazella, M. A. (1983). The use and limitations of stabilized Zirconia oxygen sensors in fluidized-bed coal combustors. In *Proceedings Electrochemical Soc (United States)* (Vol. 83). Columbus Laboratories.
57. Jordan, B. F., Baudelet, C., & Gallez, B. (1998). Carbon-centered radicals as oxygen sensors for in vivo electron paramagnetic resonance: Screening for an optimal probe among commercially available charcoals. *Magnetic Resonance Materials in Physics, Biology and Medicine, 7*(2), 121–129.
58. Qiu, X., Li, J., Wei, Y., Zhang, E., Li, N., Li, C., Yuan, H., & Zang, Z. (2019). Study on the oxidation and release of gases in spontaneous coal combustion using a dual-species sensor employing laser absorption spectroscopy. *Infrared Physics & Technology, 102*, 103042.

59. Sloss, L. L. (2011). *Efficiency and emissions monitoring and reporting.* IEA Clean Cloal Centre.
60. Yokogawa. (2008). *Carbon monoxide measurement in coal-fired power boilers.*
61. ABB. (2006). *Analytical instruments smart analyzer 90.*
62. GE. (2012). OxyTrak 390 panametrics flue gas oxygen analyzer. In *General electric measurement and control.*
63. Servomex. (2006). *Combustion analysis overview.* Servomex Group Limited.
64. Rosemount. (2010). *Model CCO 5500 carbon monoxide (CO) analyzer.* Rosemount Analytical.
65. SICK. (2013). *GM35 in-situ IR gas analyser.* SICK AG.
66. Zhang, R., Cheng, Y., Li, Y., Zhou, D., & Cheng, S. (2019). Image-based flame detection and combustion analysis for blast furnace raceway. *IEEE Transactions on Instrumentation and Measurement, 68*(4), 1120–1131.
67. Vandermeer, W. (1998). *Flame safeguard controls in multi-burner environments.* Fireye.
68. Fuller, T. A., Daw, S. C., Finney, C. E. A., Musgrove, B., Stallings, J., Flynn, T. J., Bailey, R. T., Hutchinson, D., & Lassahn, R. (2004). *Advances in utility applications of the flame doctor system.* Proceedings of the DOE-EPRI-EPAAWMA Combined Air Pollutant Control Mega Symposium.
69. Fireye. (2013). *45RM4 fiber optic flame scanner model 1001LF.* Fireye.
70. Wilcox, B. (2009). *Combustion tuning services optimize unit performance.* Babcock and Wilcox.
71. Sarunac, N., & Romero, C. E. (2003). *Sensor and control challenges for improved combustion control.* The Proceedings of the 28th International Technical Conference on Coal Utilization and Fuel Systems Clearwater, 1255. FL, USA: performance and reduced power plant emissions.
72. ABB. (2006). *Pulverised fuel flowmeter.* ABB.
73. Corp, Air Monitor. (2007). PF-FLO III. In *Air monitor power division*, 8.
74. Roberts, K., Yan, Y., & Carter, R. M. (2007). On-line sizing and velocity measurement of particles in pneumatic pipelines through digital imaging. *Journal of Physics: Conference Series, 85*, 12019.
75. Promecon, N. D., *Mecontrol air.* Promecon.
76. *Mass Flow Meter Product Guide.* (n.d.). Monterey, CA, USA: Kurz Instruments.
77. Yoder, J. (2013). The history and evolution of thermal flowmeters. *Flow Control Magazine*, 22–25.
78. EUtech. (2015). *EUtech scientific engineering power generation solutions.* EUtech Scientific Engineering.
79. Dong, Z., Wang, R., Fan, M., Xiang, F., & Geng, S. (2019). Integrated estimation model of clean coal ash content for froth flotation based on model updating and multiple LS-SVMs. *Physicochemical Problems of Mineral Processing, 55*(1), 21–37.
80. Shuk, P., McGuire, C., & Brosha, E. (2019). Methane gas sensing technologies in combustion: Comprehensive review. *Sensors & Transducers, 229* (LA-UR-18–31101).
81. Muhanji, S. O., Flint, A. E., & Farid, A. M. (2019). EIoT as a solution to energy-management change drivers. In *EIoT* (pp. 1–15). Springer.
82. Yardibi, T., Ganesh, M., & Johnson, T. L. (2018). Electrical substation fault monitoring and diagnostics. Google Patents.
83. Shaltaeva, Y. R., Podlepetsky, B. I., & Pershenkov, V. S. (2017). Detection of gas traces using semiconductor sensors, ion mobility spectrometry, and mass spectrometry. *European Journal of Mass Spectrometry, 23*(4), 217–224.
84. Dong, L., Zhang, D., Wang, T., Wang, Q., & Han, R. (2018). *On line monitoring of substation equipment temperature based on fiber Bragg grating.* 2018 2nd IEEE Advanced Information Management, Communicates, Electronic and Automation Control Conference (IMCEC), pp. 1500–1503. IEEE.
85. Kurrer, R., & Feser, K. (1998). The application of ultra-high-frequency partial discharge measurements to gas-insulated substations. *IEEE Transactions on Power Delivery, 13*(3), 777–782.
86. Hoffman, G. R. (2008). Sensing load tap changer (LTC) conditions. Google Patents.

87. Shekari, T., Bayens, C., Cohen, M., Graber, L., & Beyah, R. (2019). RFDIDS: Radio Frequency-Based Distributed Intrusion Detection System for the Power Grid. In *NDSS*.
88. Lancaster, M. (2019). *Power line maintenance monitoring*. US Patent Application No. 10/205,307.
89. Deb, S., Das, S., Pradhan, A. K., Banik, A., Chatterjee, B., & Dalai, S. (2019). Estimation of contamination level of overhead insulators based on surface leakage current employing detrended fluctuation analysis. *IEEE Transactions on Industrial Electronics, 67*(7), 5729–5736.
90. Firouzjah, K. G. (2018). Distribution network expansion based on the optimized protective distance of surge arresters. *IEEE Transactions on Power Delivery, 33*(4), 1735–1743.
91. Kuang, Y., Li, Y., Deng, Y., Huang, D., & Qiu, Z. (2019). Electric field analysis and structure design of the box of bird guard used in 220 KV transmission line. *The Journal of Engineering, 2019*(16), 2860–2863.
92. van de Kaa, G., Fens, T., Rezaei, J., Kaynak, D., Hatun, Z., & Tsilimeni-Archangelidi, A. (2019). Realizing smart meter connectivity: Analyzing the competing technologies power line communication, mobile telephony, and radio frequency using the best worst method. *Renewable and Sustainable Energy Reviews, 103*, 320–327.
93. Silva-Leon, J., Cioncolini, A., Nabawy, M. R. A., Revell, A., & Kennaugh, A. (2019). Simultaneous wind and solar energy harvesting with inverted flags. *Applied Energy, 239*, 846–858.
94. DoE, U. S. (2009). Concentrating solar power commercial application study: Reducing water consumption of concentrating solar power electricity generation. In *Report to congress*. USDOE.
95. Lapo, K. E., Hinkelman, L. M., Sumargo, E., Hughes, M., & Lundquist, J. D. (2017). A critical evaluation of modeled solar irradiance over California for hydrologic and land surface modeling. *Journal of Geophysical Research: Atmospheres, 122*(1), 299–317.
96. Banta, R. M., Pichugina, Y. L., Alan Brewer, W., James, E. P., Olson, J. B., Benjamin, S. G., Carley, J. R., Bianco, L., Djalalova, I. V., & Wilczak, J. M. (2018). Evaluating and improving NWP forecast models for the future: How the needs of offshore wind energy can point the way. *Bulletin of the American Meteorological Society, 99*(6), 1155–1176.
97. Battisti, L., & Ricci, M. (2018). *Wind energy exploitation in urban environment*. Springer.
98. Bianco, L., Friedrich, K., Wilczak, J. M., Hazen, D., Wolfe, D., Delgado, R., Oncley, S. P., & Lundquist, J. K. (2017). Assessing the accuracy of microwave radiometers and radio acoustic sounding systems for wind energy applications. *Atmospheric Measurement Techniques, 10*(5), 1707–1721.

Digital Twin Technology for Energy Management Systems to Tackle Climate Change Challenges

Eman Ahmed, M. A. Farag, Ashraf Darwish, and Aboul Ella Hassanien

1 Introduction

A digital replica of Earth can help researchers and scientists develop scientific solutions for the planet's future and model solutions to problems caused by climate change. Along with artificial intelligence, digital twinning (DT) can conduct automated monitoring of climate risks and threats, ecosystem services or threats to biodiversity, where DT can understand the challenges and choices to achieve various sustainable development goals and multilateral environmental agreements. However, there are many challenges to monitor the health of planets in real-time, including the cost of building an interoperable digital twin of the Earth and its component subsystems that can allow the monitoring, control, and modeling of complex environmental and climatic relationships.

Industrial digitalization via energy DT is regarded as a tool to efficiently optimize and manage site operations to reduce certain consumption of energy, help with energy-efficient design and evolution of their production sites and processes, and establish a green energy roadmap to switch to renewable fuels and better connect sites with locally generated renewable energy.

E. Ahmed (✉)
Faculty of Computers and Artificial Intelligence, Cairo University, Cairo, Egypt
e-mail: e.ahmed@fci-cu.edu.eg

M. A. Farag
Department of Basic Engineering Science, Faculty of Engineering, Menoufia University, Shibin El Kom, Egypt

A. Darwish
Faculty of Science, Helwan University, Cairo, Egypt

A. E. Hassanien
Faculty of Computers and Artificial Intelligence, Cairo University, Cairo, Egypt

E. Ahmed · M. A. Farag · A. Darwish · A. E. Hassanien
Scientific Research Group in Egypt (SRGE), Giza, Egypt

© The Author(s), under exclusive license to Springer Nature Switzerland AG 2023
A. E. Hassanien and A. Darwish (eds.), *The Power of Data: Driving Climate Change with Data Science and Artificial Intelligence Innovations*, Studies in Big Data 118,
https://doi.org/10.1007/978-3-031-22456-0_8

The idea of using digital twins in climate is that a physical system, process, or object can be recreated/replicated in the digital world such as carbon plants and then exposed to a range of conditions, scenarios, and risks. The impact of these exchanges in the digital world can be analyzed to understand the different results that can be achieved. Therefore, techniques for solutions can be developed to reduce carbon emissions. For example, the European Union plans to design a "digital twin of the planet" that simulates the atmosphere, oceans, ice, and land with high accuracy and provides forecasts for floods, droughts, fires, etc. In addition to capturing human behavior, enabling decision makers and governments to see the impact of climate events and climate change on society and measure the impact of various climate policies. Figure 1 depicts the publications based on the Web of Science Database from 2012 to 2022 with the citations over these years.

The interconnection of the physical world of climate with DT and metaverse technologies is a good investment opportunity as the world works from anywhere to reduce the impact of climate change [1]. Another example of the importance of using emerging technologies, including drones, to take pictures of cities to build a digital twin, as it helps decision makers or city officials to analyze the data of the captured images of trees or traffic in terms of the degree of health of trees or the degree of complexity of traffic and how they took quick decisions to provide solutions to these problems. As the volume of data is large and the degree of complexity of city infrastructure data, the role of digital twinning and artificial intelligence comes in dealing with this huge data, and building and designing software, that reduce their

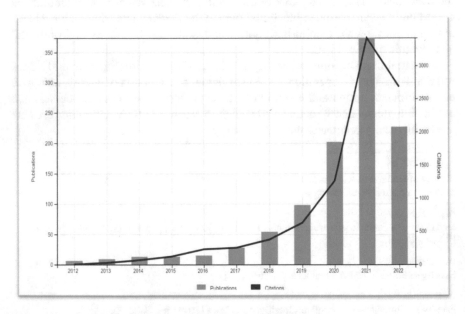

Fig. 1 Times cited and publications over time from web of science database using keywords digital twins and energy: total publications are 1039 from 2012 to 2022

carbon emissions. The application of digital twin technology to something related to the climate such as cement factories, petrochemical factories, cars, airplanes, etc. is a process of providing decision-makers with actionable insights that enable them to make improvements on things related to the climate, thus reducing carbon emissions as the digital twin of things related to it are created.

With the advent of Industry 4.0 and the rapid rise of digitalization, various technologies are advancing at different rates. Many benefits and obstacles arise with the significant improvements in DT and their accompanying technologies.

The structure of this chapter is set as follows. In Sect. 2 a discussion of the history and origin of DT is set forth, while Sect. 3 highlights the main application of DT in the energy sector. Section 4 presents a new framework for DT in the climate change scenario. Section 5 proposes a new type to DT which is called collaborative DT. Furthermore, the main problems, challenges, and potential solutions are outlined in Sect. 6. Finally, this chapter is concluded in Sect. 7.

2 Basics and Background

2.1 Digital Twins Overview

In recent years, DT has been used for industrial purposes. Moreover, it has also been utilized in some other applications like smart cities, aerospace, and healthcare. Michael Grieves proposed a prototype for a "Product Lifecycle Management (PLM) center" in 2002 [2], which included all of the DT technology's core features. As per Grieves, DT can be looked at as a virtual model of what was established. To gain a better understanding of what was created against what was designed, compare a DT to its engineering design which will fill the design-to-execution gap [3]. At first, it was primarily employed in the aerospace and astronautics fields. With the novel advances in the infrastructure technologies of the Internet of Things, there exist now a possibility to produce cheaper sensors that can collect data in real-time. Experts in electronics, sensing data collection, and the Internet of Things are involved in the data management processes when DT is implemented for the preservation and administration of historical assets.

The DT is dependent on three primary components: data and information linkages, a physical product that is in a real-world environment that is properly monitored, and a virtual product existing in a virtual space. In the first step, sensors can be used to preserve historical sites. Sensors perform monitoring in real-time and produce the huge volume of data. Following the transfer and storage of the data, it is then examined and linked to the virtual product, bringing to light information about how the physical space performs, how virtual space can be simulative, and how real-world decision-making can be done.

All the different definitions of DT lead us to differentiate between 3 similar terms:

1. The digital model is the first term, referred to as a digital version of a pre-existing or future physical object. Here, there is not any automatic data exchange between the digital model and the physical object.
2. The second term is digital shadow, which is defined as an object's digital representation with only one direction of data flow: a change in the digital object is inferred by a change in the physical item's state.
3. Finally, the digital twin is the last term. The point of contrast here is the full integration of the flow of data between the digital object and the existing physical object in the two directions

Three primary components should be included in the digital twin: the physical object, its virtual model, and the data and information connections able to achieve a link between the virtual and physical objects.

2.2 Digital Twins Origin, Concept, and Scenario

NASA first used the DT term [4]. The Apollo program proposed the DT idea, which contained the potential for building a real spacecraft to realistically replicate the original spacecraft's model. In this case, the physical vehicle that is on a space mission is synonymous with the one that stays on the ground station [5]. The work in [6] is among the first in the literature which applied DT in a study. Nonetheless, the phrase "digital equal to a physical product" was coined by Michael Grieves in 2003, widely regarded as the first time the expression DT was used [7]. DTs are (substitute) virtual models for real-world items based on virtual models and communication capabilities of services and things. [9]. Broth et al. in [8], the researchers defined it as a connected and synchronized digital replica of a real asset that describes this object's behavior. In several disciplines, the definition of the DT has been used as a virtual model to reflect the behavior of a real thing. Data that connect the digital model and the physical object, as well as the real-time bidirectional interaction between virtual representation and physical items, are deemed as the DT's essential elements.

A high number of researchers and academics have highlighted that DT is independent of specific domains. Thus, DT is ubiquitous in international industries. DT can be utilized in various ways since it is not a particular technology. Consequently, DT requires a definition that is more precise and industry-specific.

Model-based simulation technologies, a variety of sensors, the Internet of Things, and modeling tools are all used in DT applications. RFID tags, gauges, cameras, scanners, readers, and other data-related devices are examples of such devices. Massive amounts of unstructured, semi-structured, and structured data are routinely produced by them. Edge computing is utilized for pre-processing of the data obtained since it is expensive and complex to transmit these data to the DT in the cloud server. 5G/6G technology removes the data-leakage threat and makes sure the data is transmitted in real-time. DT models are divided into two types: data models and physical models. Physical models necessitate an understanding of physical properties. Meanwhile, semantic data models must be educated using Artificial Intelligence techniques using specific output and input data. When using model-based simulation technologies simulation becomes a DT feature that cannot be avoided. The DT simulation lets the physical model conduct real-time interaction with the virtual model.

2.3 Digital Twin's Characteristics

A DT technology controls the process of physical space. The traditional job of DT is to monitor scattered devices in real-time across an IoT network. Data representation for machine-readable application programming interfaces Another critical role of DT is the implementation of a logic function for event detection.

Terms such as "digital shadow," "digital model," and "digital twin" may be employed while working with DT technology. Each of these terms has a distinct meaning that is not found elsewhere. Among them are the following:

(a) Digital shadow is a digital replica of an object with data flowing from a physical object to the virtual representation in one direction.
(b) A digital model is a digital representation or replica of a real-world or planned thing. The physical thing and the model are unable to communicate with one another. This category will cover items such as product designs, building plans, and so on.
(c) A digital twin exists when data flows between an existing physical thing and a digital object. Any change to the physical object impacts the digital object and vice versa.

2.4 Internet of Things

To collect data from a physical twin, in reality, a system for manipulating sensor nodes is required.

The use of IoT in a variety of technical and scientific fields, such as transportation, smart environments, manufacturing, healthcare, and other fields.

Because these applications create enormous volumes of data, a data analysis system is required for system fault detection maintenance and prediction. A wide number of application domains are also available through the IoT, including those for the automotive, traffic management, medical assistance, mobile healthcare, geriatric support, intelligent energy management, home automation, and industrial automation industries. IoT technology continues to enable a wide range of applications, including decision support systems, analytics, and real-time monitoring. IoT and other distributed generation technologies are linked to digital twins.

2.5   Sensor Networks

Applications for wireless sensor networks (WSN) can be found in several sectors, including industry and healthcare. WSN is used in a range of industrial and healthcare applications, such as gas detection and indoor/outdoor environmental monitoring. Environmental parameters such as temperature and humidity have been managed and monitored using WSN. The presence of low-cost micro-scale sensing devices has increased the possibilities for the creation of WSNs, which can now incorporate new and complicated environmental parameters such as airflow, light intensity, temperature, and contaminants.

WSN is widely used in the energy sector. A WSN monitoring system deployment offers a variety of benefits, including architecture scalability, the integration of several heterogeneous sensors on a single compact node, and the distribution of a large number of wireless and reasonably priced measurement points. Therefore, WSN must be incorporated into DT applications to significantly improve the process of preserving the energy sector.

3   Digital Twins in Energy Management Applications

The DT can be split into various layers in energy applications, based on the integration level. The main application of DT in the energy sector is described in Fig. 2.

(1)   Low carbon emissions

After the assessment of cities by many indicators and criteria, it may be concluded that in a low-carbon city there is an overall lower carbon emissions level compared to the threshold when weighing the carbon emissions of each criterion. In low-carbon cities, there is a higher level of control of carbon dioxide emissions. Four categories contribute to emissions: urban transportation, structures, environments,

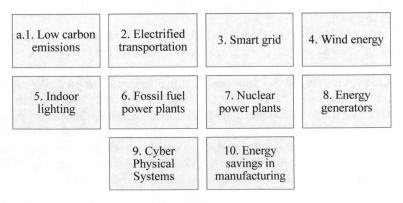

Fig. 2 Digital twins in energy management applications

and infrastructure. Low carbon cities were created as a response to the increasing need to reduce carbon emissions and lessen the effects of climate change in urban areas [10, 11]. In this context, DT is frequently used to simulate complex systems, which correlates to the capabilities of DT in high-precision modeling. This cover also a variety of different applications like smart charging management, urban energy planning, and improving energy efficiency. While preserving or increasing the level of well-known economic activity, the low-carbon city works to minimize energy use.

The benefits of DT and pertinent technology advancements used in low-carbon cities are outlined below:

1. Making sure low carbon production is reduced as much as possible.
2. Warning about behaviors of high emission.
3. Decreasing the economy-generated carbon emissions.
4. Testing the energy strategies online.
5. Monitoring unusual consumption of energy.

(2) Electrified transportation

Electric traction motors are typically used in electric cars to provide propulsion. In an effective energy storage system, ultra-capacitors, and Li-ion batteries, for example, can be charged by solar energy, which powers these motors. As well as, in hybrid automobiles, the use of electric motors and batteries enables internal combustion engines to function more efficiently and emit fewer greenhouse pollutants.

In light transportation, battery-electric vehicles already possess a significantly larger decolonization potential than short-distance vehicles powered by gasoline and biofuels. This permits the use of DT in a variety of applications, hastening the development of electric transportation.

Even more extensive and varied are the uses of DT in electro-powered transport, yet only a small number of these can be connected to the smart grid.

The benefits of DT and associated technology advancements used in electrified transportation are listed below:

1. Compatibility with path planning algorithms
2. Better portability of environment modeling
3. Decreased pressure on cloud computing
4. Real-time optimization of traffic energy usage as a part of a higher level: data collecting and global optimization.

(3) Smart grid

The term "smart grid" refers to a grid that makes use of computer technology to enhance connectivity, automation, and communication among energy networks. Smart grids already leverage technologies like big data processing, cloud computing, and reinforcement learning. Smart grids must also incorporate common nonrenewable and renewable energy sources to decrease environmental risks and improve sustainability [12–14].

Different parts of the smart grid can make use of DT. Its ability to track and reflect the full life cycle is the reason behind this. DT may be able to use some of the applications below to better improve and build upon its strengths. Building a model of the architecture of a smart grid, aggregating data in a smart grid, designing smart meters, and analyzing smart meter data are a few examples.

To sum up, the most pertinent literature shows that the framework of Industry 4.0 has significantly contributed to enhancing advanced simulation technologies and techniques, like DT. In addition, power distribution networks are incorporating DTs as well, making it simpler to regulate and manage the energy supply network.

(4) **Digital twins in the wind energy sector**

In the wind energy sector, the development of DT as a cutting-edge technology has made it possible to design, control, and forecast the performance of wind turbines. The expert decisions are examined over what to leave out and include in the design of DT in understanding twinning as a collection of dynamic design processes that have implications for how wind energy is handled and developed. The twinning is considered a process of governance by design, where twinning decisions may steer wind power advancements and many industry participants' decisions and actions [15].

(5) **Digital Twins for indoor lighting**

Enhancing the energy conversion efficiency of lighting systems has become a critical study field of green building energy conservation within the standards for assuring lighting quality and illuminance scale. Sensor control and LED lights are currently the most commonly utilized energy-saving technologies for lighting systems, replacing manual switch control and fluorescent lights. These strategies can significantly reduce lighting system energy usage, but they result in a significant rise in initial installation costs. The incapacity of modern lighting systems to implement integrated and intelligent management presents huge hurdles for the whole life cycle of operation, and maintenance (O&M) and energy consumption forecast.

To solve the inadequacies of current lighting systems, more focus has been placed on integrating indoor intelligent lighting systems. To perform more complicated intelligent control and improve the ability of lighting systems to communicate with their surroundings, several researchers have started to link sensors and the Internet of Things (IoT). Since its introduction, DT has gained prominence and is now seen as a crucial enabler of the shift to Industry 4.0. A DT can represent an actual built environment in a virtual setting and replicate the interconnected processes that occur throughout the environment's whole life cycle in real-time.

Currently, several energy-saving studies focusing on light sources and control systems have been carried out with great outcomes. For instance, utilizing new LED lights and sensor control both reduced lighting energy use by 10–25% and by more than 50%, respectively [16].

(6) Digital twin for fossil fuel power plants

The great opportunities that DT applications for power plants provide are currently being pursued by the energy sector. Digitalization and connected plant technologies that leverage DTs for rapid power system transition and to minimize the impact of the plant cycle can improve the operational flexibility of power plants. However, numerous critical DT components require additional research and development. One of these components is the sensor network architecture, which may be adjusted utilizing a two-tier approach: component data and plant for performance optimization, condition monitoring, and problem diagnostics. Virtual reality (VR) technologies and integrated dynamic simulation (IDS), enhanced process control, and methods for flexible power plant management are the other essential components of DT that require further development.

(7) Digital twin for nuclear power plants

The DT uses a network of interconnected multi-scale, multi-physics models to replicate real-world conditions such as inspections and in-service monitoring, post-shutdown inspections, and in-situ waste storage monitoring. The planned DT's implementation, shortcomings, and benefits are identified and explored, with an emphasis on future developments in supercomputing, building algorithms for handling massive amounts of data, and the significance of acquiring data through measurement innovation, uncertainty, and analysis. Applications of DT are seen in [17] for nuclear power reactors.

(8) DT for renewable energy generators

The DT has the potential to play a significant role in the optimal design and dependable operation of large renewable energy systems. The authors of [18] emphasized the importance and limitations of DT models for large renewable energy sources. They presented a multi-domain live simulation platform for wind and hydropower plants, as well as a thorough modeling technique.

(9) DT for energy cyber-physical systems

The approach to modeling energy cyber-physical systems (ECPS), which was first introduced by [19], has been adopted by the energy sector. The authors in [20] presented DT and intelligent DT architectures for cyber-physical production systems. The implementation and assessment were conducted using an agent-based approach for DT simulations along with a heterogeneous data gathering and integration mechanism. For a use case involving metal forging, they presented intelligent DTs that are partially realized. Future studies should address why this realization for energy cyber-physical systems was not implemented.

(10) DT for energy savings in manufacturing

A recent evaluation of DTs for industrial energy savings applications by [21] explored at the possibilities for a more precise and effective DT-based infrastructure. The researchers suggested standardizing and modularizing industrial data infrastructure to adopt advanced energy-saving solutions. They also offered implementation guidelines.

4 A Proposed Framework of Digital Twins in Climate Change Adaptation

Climate change occurs due to natural processes affected by the changes in the environment and human intervention as shown in Fig. 3. Climate change is inevitable. Accordingly, we have to cope with it by finding solutions that achieve human needs and adapt to the variations.

DT can play the main role in climate change adaptation. Destination Earth is a European Union initiative to create a digital model of Earth that will be used to monitor the effects of natural and human activity on our planet, it focuses on combining digital twins to create a complete replica of the systems of Earth over the next decade [22]. Digital twins role in the greenhouse is discussed in [23].

Human needs arise from different socio-economic sectors such as health, food and agriculture, hydrology and water needs, economics and transport. Dt can be

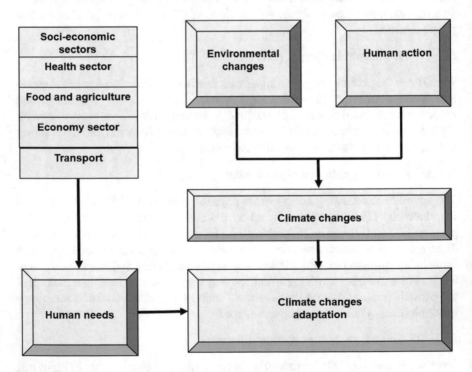

Fig. 3 Climate change process

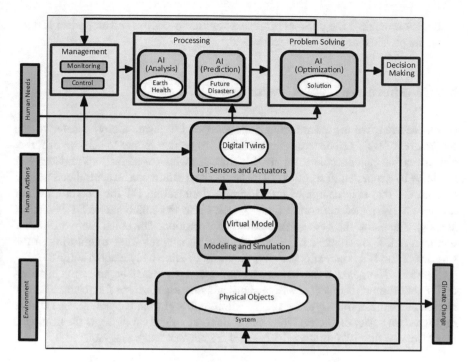

Fig. 4 The proposed framework for using digital twins in a climate change scenario

used to monitor and analyze the current situation and predict any future disasters, then, find a solution that compromises climate change and human needs. This is done by applying AI techniques to DT. First, the proposed solution is tested on the digital twins for assessment, this is done on all the solutions suggested by the problem-solving module. Then, a decision is made by choosing a solution and applying it in the real world. Figure 4 presents the proposed framework.

The inputs are the environment and the human actions, the output is the resulting climate change. The system's physical objects are converted into digital twins using modeling and simulation, and IoT sensors and actuators that provide real-time feedback on the statuses.

Four main operations can be accomplished using digital twins:

- Management: It is responsible for monitoring the statuses of the DT and controlling them.
- Processing: It uses AI techniques to analyze the earth's health regularly and predict any possible future disasters.
- Problem-solving: It takes into account the human needs in addition to the analyzed current situation and gets a possible solution that achieves the human needs with minimal impact on the environment. The solution is then reflected on the digital twins for assessment.

- Decision making: the optimal proposed solution is chosen and an update is done on the physical objects in the real world.

5   Collaborative Digital Twins

In this section, we are introducing collaborative DT terminology to refer to the integration of DT from different organizations to work together to achieve different objectives that benefit society. A network is used in which organizations register their DT to be incorporated in solutions. The network structure can be centralized having a controller that is in charge of monitoring and allocating DT for a certain task or it can be a distributed network. Figure 5 shows how the collaborative DT is formed and Fig. 7 explains the general framework of operation. The collaborative DT uses its member DT to achieve a required objective. Having real-time monitoring of the statuses of the DT, it can allocate an optimal set to effectively reach a solution.

There are N organizations denoted by $O_1, O_2, ..., O_N$. Each organization has a set of DT that mimic its physical assets statuses using input sensor feedback. The DT has features corresponding to the specifications of the physical object. These DTs are registered on a network so that they can collaborate. Figure 6 presents the proposed framework of the collaborative DT general framework block diagram.

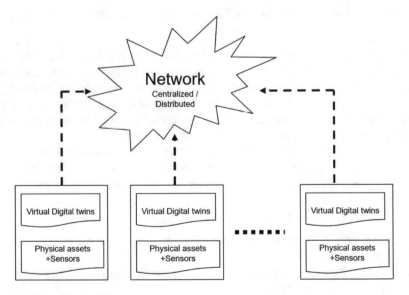

Fig. 5 Collaborative digital twins

Fig. 6 Collaborative digital
twins general framework
block diagram

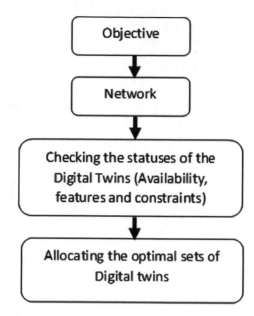

Given an objective, the network checks the statuses of the monitored DT on the network. Then, it assigns the optimal set of the DT to achieve this objective. The collaborative DT is responsible for:

1. having a record of all the network DT.
2. monitoring the status of each of the DT in the network.
3. checking the availability of any DT in the network at any time.
4. predicting maintenance requirements for the DT.
5. predicting any possible future failure in the DT to alleviate them.
6. allocating of DT to different tasks.
7. optimizing the usage of DT.

5.1 The Need for Collaborative Digital Twins

DT technology opens great insights of collaborations between different organizations to achieve greater objectives that benefit the society and the environment rather than a single organization. Objectives may include reduction of carbon dioxide emission and energy minimization. Such objectives that serve the environment can't be obtained without sharing of resources and optimizing their usage. Collaborative DT enables the optimal usage and management of DT from several organizations to reach a certain purpose.

Fig. 7 Block diagram of
using the collaborative
digital twins in the
healthcare use case

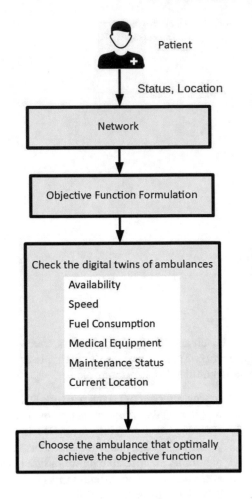

5.2   Use Case of Collaborative Digital Twins in Healthcare

In the healthcare sector, collaborative DT help in minimizing the risk of losing lives and getting to the patient in an emergency as soon as possible. Assuming that different hospitals join a collaborative DT network with their ambulance's DT. Given the status of a patient in an emergency, an input sensor sends a signal to the network of collaborative DT with the patient's health condition and his location. The network checks the availability of ambulances in the near hospitals, their locations, and their statuses. It then decides which ambulance to send according to the shortest and fastest route between the patient location and the ambulance location. Figure 7 illustrates the whole procedure. This setting minimizes time and the risk of losing the patient. This is in addition to minimizing energy consumption and carbon dioxide emitted by cars. It is to be noted that collaborative DT requires the collaboration of different hospitals to share their data about the digital twins of their ambulances. It could be applied using a single hospital, but this won't be as effective as having several hospitals. This can be noted in case that either the hospital location or the ambulances' locations are far from the patient, or the number of its ambulances is limited and none is available at the time of emergency. Increasing the number of organizations in the network means increasing the shared DT resources, which in turn enables us to get better optimization.

6   Challenges and Potential Solutions

6.1   *Problems and Challenges*

Recently, the capacity of DT technology to merge physical and virtual space has attracted a lot of interest and it has potential for a wide range of uses. However, there are currently some issues with DT technology [24]. The issues that must be addressed are summarized below:

1. It is difficult for different DT models to communicate with one another. Data's inconsistent semantic syntax causes redundant or missing knowledge resources, as well as impediments to basic knowledge base interoperability between levels. It is difficult for several DT models to work together. Although there is a clear connection between knowledge and other concepts, it is difficult to explain because each data source's structure varies. To build a knowledge graph and add value, it is becoming more and more crucial to choose valuable knowledge. Data processing from a single system is the primary emphasis of conventional goods and solutions. However, when there is a vast and complicated quantity of data that has to be processed, it must be solved by applying semantic engineering techniques to construct knowledge graphs. Structured data is processed using traditional solutions. Unstructured data still needs to be formatted for specific contexts before processing, and knowledge graphs are the only tools that can achieve this.
2. Developing DT models repeatedly for many domains and application situations is laborious and time-consuming. Problems with model and data interchange will arise when DT models are moved to other platforms. DT-related data sharing mechanisms and service systems are still being developed, and sharing data and models across different topics poses security issues and potential conflicts of interest, making it challenging to adhere to the necessary DT standards for data generation and sharing.
3. The standard DT paradigm is incapable of converting external input data into conceptual, logical data. The model cannot directly operate on the majority of the data collected. To obtain specific relevant model parameters or operational data, the data must be interpreted. Effective real-time computer capabilities are required for this process. Furthermore, during iterative optimization, the DT

model is ineffectual in logical judgment and self-determination. This makes designing and finishing the DT model more complex.

4. DT activities necessitate the processing of huge volumes of heterogeneous data, which is a more severe test of communication and storage capability. Fault diagnosis and preventative maintenance of complex goods, for example, need a massive data storage and intensive data analysis from numerous data sources, offering a bigger challenge to data storage capacity and processing capabilities. These sorts of digital twin application scenarios prioritize enormous data storage and processing above rapid data processing.

5. Because there are no standards for DT design, development, management, and operation, DT is difficult to copy, learn, and imitate. Various types of data transfer and interaction between different systems and devices are necessary during the process of developing and integrating DT models. As a result, developing data standards and communication interface protocols, as well as unifying data semantics and codes, is crucial for constructing a comprehensive multidimensional DT platform. The lack of homogeneity in existing communication interface protocols and data standards for many systems and devices is a significant barrier to DT development.

6. The DT model often consists of a mechanism or decision model that does not provide feedback or updates for FLC management. Existing DT models cannot accurately anticipate product life and manage dynamic operational processes after manufacture. Additionally, given the DT model's major objective, monitoring the product's health and extending its life should be the key focus. Although there are significant gaps in the production and reuse stages, existing DT models are typically applied to the product's functioning period. Using FLC data to update and optimize the model state and parameters can result in an adaptive DT model.

6.2 The Potential Solutions

6.2.1 Knowledge Graph Analysis

The set of rules used in the creation of an expert system is called a knowledge base, which also includes the information and facts related to the rules. A knowledge graph is a collection of many graphs illustrating the connection between the knowledge structure and the development process. It describes knowledge resources and their carriers using visualization technology, and it draws, constructs, mines, and displays knowledge as well as the connections between it. Many different degrees of expertise are required to establish a digital twin model, and those databases are not yet ideal. Redundant or missing knowledge is produced by the current difficulties in the interoperability of various models and the absence of common data semantics.. One possible option could be to build a good knowledge graph and harmonies data semantics.

6.2.2 DT Models Migration

The reuse of distinct models from various domains or scenarios inside the same domain is referred to as model migration. Model migration can minimize modeling complexity, accelerate model construction, and broaden model application in a variety of operating environments. After the modeling work is finished, model migration in many situations is considered for DT models with high accuracy. Consider linking DT models in multiple domains using completed models versus fresh models, which is illuminating for model construction.

6.2.3 AI Technologies

DT is a complete solution that connects with numerous AI technologies, including cyber-physical systems (CPS), IoT, and machine learning. Real-time computing is an issue for DT that must be taken into consideration and AI greatly aids in this area. It may effectively lower the difficulty of model creation and increase efficiency when used with machine learning. The dynamic control and real-time sensing of engineering systems, which are crucial in DT, are therefore made possible by CPS and IoT.

6.2.4 Data Processing

For DT, the capacity to store and process more data is essential. Accuracy, data storage, transmission stability, and consistency are subsequent components of data processing. Furthermore, at various phases of model usage, numerous sources of heterogeneous data must be merged and fused to generate a single data carrier. A cloud platform is widely utilized to handle complicated data storage problems. Enhancements and breakthroughs in communication technologies will help to increase the stability, security, and dependability of data transfer. Data standards and communication protocols can handle the integration and fusion of disparate data sets. Additionally, as data is the foundation of manufacturing service collaboration in DT, both the added value of the model and the value of the data must be assessed. This problem is thought to be addressable by enhancing the value chain of DT models and developing an industrial chain in DT.

6.2.5 Standardization

A standardized framework that takes into account the interplay of platforms, software, interfaces, and technical rule coordination is necessary for the DT. The DT's general standardization effort is also in its early stages, and standard research content must be supplemented. Standardization content that may be addressed involves standards for fundamental aspects, concepts, technological implementation, testing, and assessment, as well as standards for collaboration across different systems.

6.2.6   Digital First

At this point, we might think of DT as a digital parallel stage in which the virtual digital, and physical worlds are mapped and viewed concurrently. Digital-first refers to the process of generating digital virtual before real manufacturing to actively think, deduce, and discuss. There is a possibility to conduct digital first when adding a significant quantity of external or historical data based on DT and exploiting the high processing capability of computer equipment. Following that, several repetitions of low-cost studies could save significant capital expenses and time. This could be an innovative technique to solve complex challenges in the future. generation.

7   Concluding Remarks

The future trend of energy technology will be the deep integration of networking, digitization, and intelligence with energy applications, as science and technology improve and explore. The maturation of digital technologies such as Big Data, the Internet of Things, and cloud computing has resulted in the development and application of a prospective technology Digital Twin. Many concepts and outcomes from Digital Twin research are already being used in a variety of sectors.

The current chapter describes DT technology and its applications in the energy area, as well as essential technologies. First, the development history and research background of DT are explained, followed by a discussion of the important and widespread uses of DT in the energy industry, particularly in power-related systems. For the first time, several levels of integration are recommended for DT. The smart energy application sector is divided into four categories: smart grid, electrified transportation, low carbon city, and advanced energy storage system. Each level is discussed through literary examples.

With this discussion and review, the future issues that DT will face, as well as the path of technology development, are addressed.

References

1. Kite-Powell, J. (2022). *Can digital twins drive a climate change agenda?* May 16, 2022. https://www.forbes.com/sites/jenniferhicks

2. Grieves, M., & Vickers, J. (2016). Digital twin: Mitigating unpredictable, undesirable emergent behavior in complex systems. In F. J. Kahlen, S. Flumerfelt, & A. Alves (Eds.), *Transdisciplinary perspectives on complex systems: New findings and approaches* (pp. 85–113). https://doi.org/10.1007/978-3-319-38756-7_4

3. Grieves, M. (2014). *Digital twin: Manufacturing excellence through virtual factory replication.* White paper, Florida Institute of Technology (pp. 1–7).

4. Shafto, M., Conroy, M., Doyle, R., Glaessgen, E., Kemp, C., & LeMoigne, J., et al. (2012). *NASA technology roadmap: Modeling, simulation, information technology & processing roadmap, technology area* (Vol. 11). National Aeronautics and Space Administration.

5. Boschert, S., & Rosen, R. (2016). Digital twin—The simulation aspect. In *Mechatronic futures* (pp. 59–74). Springer.

6. Hernandez, L., & Hernandez, S. (1997). Application of digital 3D models on urban planning and highway design. *WIT Transactions on the Built Environment, 33.*

7. Ghosh, A.K., AMM, S. U., Kubo, A. (2019). Hidden Markov model-based digital twin construction for futuristic manufacturing systems. *Artificial Intelligence for Engineering Design, Analysis and Manufacturing: AI EDAM, 33*(3), 317–331.

8. Borth, M., Verriet, J., & Muller, G. (2019). Digital twin strategies for SoS 4 challenges and 4 architecture setups for digital twins of SoS. In *2019 14th annual conference system of systems engineering (SoSE)* (pp. 164–169). https://doi.org/10.1109/SYSOSE.2019.8753860

9. Schluse, M., & Rossmann, J. (2016). From simulation to experimentable digital twins: Simulation-based development and operation of complex technical systems. In *2016 IEEE international symposium on systems engineering (ISSE)* (pp. 1–6). https://doi.org/10.1109/SysEng.2016.7753162

10. Falahatkar, S., & Rezaei, F. (2020). Towards low carbon cities: Spatio-temporal dynamics of urban form and carbon dioxide emissions. *Remote Sensing Applications: Society and Environment, 18,* 100–317. ISSN 2352-9385.https://doi.org/10.1016/j.rsase.2020.100317

11. Tan, S., Yang, J., Yan, J., Lee, C., Hashim, H., & Chen, B. (2017). A holistic low carbon city indicator framework for sustainable development. *Applied Energy, 185*(Part 2), 1919–1930. ISSN 0306-2619. https://doi.org/10.1016/j.apenergy.2016.03.041

12. Priyadarshini, I., Kumar, R., Sharma, R., Singh, P. K., & Satapathy, S. C. (2021). Identifying cyber insecurities in trustworthy space and energy sector for smart grids. *Computers and Electrical Engineering, 93,* 107–204. ISSN 0045-7906.https://doi.org/10.1016/j.compeleceng.2021.107204

13. Neffati, O. S., Sengan, S., Thangavelu, K. D., Kumar, S. D., Setiawan, R., Elangovan, M., Mani, D., & Velayutham, P. (2021). Migrating from traditional grid to smart grid in smart cities promoted in developing country. *Sustainable Energy Technologies and Assessments, 45,* 101–125. https://doi.org/10.1016/j.seta.2021.101125

14. Khalil, M. I., Jhanjhi, N., Humayun, M., Sivanesan, S., Masud, M., & Hossain, M. S. (2021). Hybrid smart grid with sustainable energy efficient resources for smart cities. *Sustainable Energy Technologies and Assessments, 46,* 101–211. https://doi.org/10.1016/j.seta.2021.101211

15. Jasanoff, S. (2016). *The ethics of invention: Technology and the human future.* WW Norton & Company.

16. Chew, I., Kalavally, V., Oo, N. W., & Parkkinen, J. (2016). Design of an energy-saving controller for an intelligent LED lighting system. *Energy Build, 120,* 1–9. https://doi.org/10.1016/j.enbuild.2016.03.041

17. Okita, T., Kawabata, T., Murayama, H., Nishino, N., & Aichi, M. (2019). A new concept of digital twin of artifact systems: Synthesizing monitoring/inspections, physical/numerical models, 935 and social system models. *Procedia CIRP, 79,* 667–672. https://doi.org/10.1016/j.procir.2019.02.048

18. Ebrahimi, A. (2019). Challenges of developing a digital twin model of renewable energy generators. In *IEEE 28th international symposium on industrial electronics (ISIE)* (pp. 1059–1066). https://doi.org/10.1109/ISIE.2019.8781529

19. Saad, A., Faddel, S., & Mohammed, O. (2020). IoT-based digital twin for energy cyber-physical systems: design and implementation. *Energies, 13*(18), 4762. ISSN 1996-1073.https://doi.org/10.3390/en13184762

20. Ashtari Talkhestani, B., Jung, T., Lindemann, B., Sahlab, N., Jazdi, N., Schloegl, W., & Weyrich, M. (2019). An architecture of an intelligent digital twin in a cyber-physical production system. *at-Automatisierungstechnik, 67*(9), 762–782. https://doi.org/10.1515/auto-2019-0039

21. Teng, S. Y., Touš, M., Leong, W. D., How, B. S., Lam, H. L., & Máša, V. (2021). Recent advances on industrial data-driven energy savings: Digital twins and infrastructures. *Renewable and Sustainable Energy Reviews, 135*, 110–208. https://doi.org/10.1016/j.rser.2020.110208

22. Ariesen-Verschuur, N., Verdouw, C., & Tekinerdogan, B. (2022). Digital twins in greenhouse horticulture: A review. *Computers and Electronics in Agriculture, 199*, 107–183. https://doi.org/10.1016/j.compag.2022.107183

23. Jia, G., et al. (2020). Digital earth for climate change research. In H. Guo, M. F. Goodchild, & A. Annoni (Eds.), *Manual of digital earth* (Vol. 11, pp. 473–494). Springer. https://doi.org/10.1007/978-981-32-9915-3_14

24. Wang, Y., Kang, X., & Chen, Z. (2022). A survey of digital twin techniques in smart manufacturing and management of energy applications. *Green Energy and Intelligent Transportation*, 100014. ISSN 2773-1537.https://doi.org/10.1016/j.geits.2022.100014

The Role of Internet of Things in Mitigating the Effect of Climate Change: Case Study: An Ozone Prediction Model

Lobna M. Abou El-Magd, Aboul Ella Hassnien, and Ashraf Darwish

1 Introduction

The term climate change refers to the alterations in the statistical distribution of the weather over extended periods, which can commonly range from decades to millions of years. These deviations can happen in the weather on average or only in how weather occurrences are distributed around an average. They might be localized to a certain place or happen everywhere in the world [1].

Both climate change and air pollution are significantly exacerbated by greenhouse gas emissions produced during the extraction and combustion of fossil fuels and account for a significant portion of both of these factors [2]. Using fossil fuels for energy, deforestation, and degradation of forests to release greenhouse gases into the atmosphere are two of the most significant human-caused contributors to climate change. Human activities have been connected to a rise in GHGs in the atmosphere, including Carbon dioxide (CO_2), methane (CH_4), nitrous oxides (N_2O), and fluorinated gases [3]. This is to inform you that as the Institutional email address of the corresponding author is not available in the manuscript, we are displaying the private

L. M. Abou El-Magd (✉)
Department of Computer Science, Misr Higher Institute for Commerce and Computers, Mansoura, Egypt
e-mail: lobna_acd@hotmail.com
URL: http://www.egyptscience.net

A. E. Hassnien
Faculty of Computers and Information, Cairo University, Giza, Egypt
URL: http://www.egyptscience.net

A. Darwish
Faculty of Science, Helwan University, Helwan, Egypt
URL: http://www.egyptscience.net

L. M. Abou El-Magd · A. E. Hassnien · A. Darwish
Scientific Research Group in Egypt, Giza, Egypt

A. E. Hassanien and A. Darwish (eds.), *The Power of Data: Driving Climate Change with Data Science and Artificial Intelligence Innovations*, Studies in Big Data 118, https://doi.org/10.1007/978-3-031-22456-0_9

email address in the PDF and SpringerLink. Do you agree with the inclusion of your private e-mail address in the final publication?

Numerous factors, including ocean salinity, oceanic CO_2 emissions, land dynamics, cloud characteristics, air temperature, and lighting, impact the climate of the entire planet. Therefore, keeping an eye on these variables is necessary if one wants to comprehend the changes occurring in the natural surroundings.

Climate change has many different effects on people's lives and health. It threatens the things most important for good health, like clean air, safe water to drink, nutritious food, and a safe place to live, and it could undo decades of progress in global health. Climate change is estimated to cause around 250,000 more fatalities annually between 2030 and 2050 due to hunger, malaria, diarrhea, and heat stress alone. Direct health-related expenses are anticipated to reach $2–4 billion annually by 2030. Areas with poor health infrastructure, primarily in poorer countries, will be the least equipped to prepare and respond without aid [4].

There were 315 natural disasters worldwide in 2018, and most of them had a climatic component. A total of 68.5 million people were impacted, and there were $131.7 billion in economic damages, of which storms, floods, wildfires, and droughts were responsible for 93% [5].

Climate change also has effects directly on human survival. For example:

a. Climate change is currently regarded as a major danger to the quality and quantity of the world food supply. The expected effects of climate change pose a growing danger to food security, particularly access to protein-rich foods [6].
b. In aquaculture, rising temperatures, shifting precipitation patterns, and increased frequency of extreme events are currently obvious on water resources, while others are emerging. Due to aquaculture's relevance to global food security, nutrition, and livelihoods [6].
c. Most systematic reviews indicate that climate change is associated with poor human health. And that there are three common diseases (1) infectious diseases (2) mortality, and (3) respiratory, cardiovascular, or neurological consequences that may occur due to climate change [7].

Traditional mitigation strategies that use decarbonization technologies and methods to lower carbon dioxide emissions, such as renewable energy, fuel substitutions, efficiency improvements, nuclear power, and carbon storage and utilization, are one strategy used to combat climate change [8]. There are activities for reducing greenhouse gas emissions to mitigate climate change, such as; designing buildings to make them more energy efficient, embracing renewable energy sources like solar, wind, and small hydro, helping cities create sustainable transport like BRT, electric vehicles, and biofuels, and supporting sustainable land and forest.

Smart cities facilitate three crucial pillars for the nation and the community: intelligence, safety, and sustainability. The level of emphasis placed on each of the pillars varies in every country based on its requirements and priorities. Africa has a tremendous possibility to grow towards the notion of smart cities. The smart city is one of the primary focuses of their ambitions in Africa. Multiple scales and degrees of national and commercial initiatives for smart cities exist in Africa. The new administrative capital of Egypt serves as a model for implementing the smart city model, among the numerous other smart city efforts now ongoing in the country. We observe many smart city efforts in South Africa, Nigeria, Rwanda, and Egypt [9].

The Internet of Things has unique prospects for solving numerous environmental challenges, such as clean water, landfill trash, deforestation, and air pollution. It will ultimately aid in reducing the environmental impacts of human activities [10]. Therefore, this study aims to discuss the role of the Internet of Things technology in reducing the negative consequences of climate change. We build an intelligent model based on IoT technology for predicting the Ozone.

The rest of the paper is organized as follows. Section 2 focuses on Preliminaries. Section 3 presents the Dataset description. The proposed IoT-based ML for building an Ozone prediction model presents in Sect. 4. The results and discussion are presented in Sect. 5. Finally, the conclusion presents in Sect. 6.

2 Preliminaries

2.1 Internet of Things

Every connected device is regarded as a thing in the context of the IoT. Things typically include physical actuators, sensors, and a microprocessor-equipped embedded system. Machine-to-Machine (M2M) communication is required since objects must interact. Wireless technologies including Wi-Fi, Bluetooth, and ZigBee can be used for short-range communication, as well as mobile networks like WiMAX, LoRa, Sigfox, CAT M1, NB-IoT, GSM, GPRS, 3G, 4G, LTE, and 5G for long-range communication [11]. IoT can assist in making many processes more quantifiable and measurable by collecting and analyzing massive amounts of data [12].

IoT has the potential to improve the quality of life in a variety of sectors, including healthcare, smart cities, the construction industry, agriculture, water management, and the energy sector [13]. It is widely utilized in environmental monitoring, healthcare systems and services, energy-efficient building management, and drone-based service delivery [14, 15].

IoT has unique prospects for solving numerous environmental challenges, such as clean water, landfill trash, deforestation, and air pollution, and will ultimately aid in reducing the environmental impacts of human activities. Renewable sources like solar and wind, which don't emit CO_2, must be used to generate electricity in the future. With IoT, consumer items can be made faster, more powerful, and more efficient while using less energy [16].

2.2 The Role of the Internet of Things on Climate Change

The introduction discussed the causes, consequences, and mitigation strategies for climate change. This section explains how to use IoT to mitigate climate change's effects. According to a paper that Ericsson just released, the Internet of Things will be responsible for a decrease in greenhouse gas emissions of 63.5 gigatons by the year 2030 [17]. This is partly due to chances for industry collaboration that wasn't possible before the IoT.

The use of information and communication technologies (ICTs) is extremely helpful in the process of monitoring climate variables. The World Meteorological Organization (WMO) and the International Telecommunication Union (ITU) emphasize the potential of information and communications technologies for the monitoring of climate change [18]. Information on climate change is mostly obtained via ICTs such as satellite-mounted sensing instruments, ocean-based sensors, and weather RADARs.

We can summarize the role of the IoT for climate mitigation in three main Categories; monitoring prediction and climate change reduction as follows:

A. *Monitoring*

According to recent research, employing the Internet of things can help in monitoring global climate change [18]. The Internet of Things paradigm is ideal for creating data, making it the ideal solution for gathering climate data. IoT devices can provide more robust data from a wider variety of sources than are now available from existing temperature, humidity, and precipitation sensors worldwide. Ocean temperature and level can now be monitored with greater precision thanks to these gadgets, and this information will be invaluable in the fight against climate change. Scientists worldwide will be able to update their models based on the precision of IoT devices, which can detect even the tiniest changes.

Several studies demonstrate that the utilization of Internet of Things technologies, such as Internet of Underwater Things (IoT), Internet of Underground Things (IoUGT), and Internet of Space Things (IIoT, can be of considerable assistance in the research of these crucial climate-altering variables [18]. The researchers in [18] suggested a framework named X-IoT. The X-IoT framework comprises intelligent sensors located in oceans, underground, and outer space. For instance, ocean sensors measure various factors, including salinity, acidity, and temperature. Similarly, buried sensors can provide information regarding subterranean soil changes, seismic activity, and gas sensing. Similarly, satellites outfitted with sensors provide information such as ocean altitude, cloud characteristics, amount of solar radiation, spatial–temporal knowledge of lands, and ocean CO_2 emission. They believe combining all of this sensing information could represent a revolutionary step forward in climate monitoring.

B. *Prediction*

The main role of IoT is to collect data. With the help of machine learning techniques, IoT can be used to predicting disasters and fight climate change in many fields [19].

i. Agriculture: The use of sensors to collect and transmit data in agriculture will result in improved precision agriculture techniques that minimize the usage of pesticides, fertilizers, and water [20]. IoT devices will improve the accuracy of weather forecasts, allowing farmers to use their resources more efficiently and reduce waste [20, 21].

ii. Utilities: Utilities can employ IoT to create a responsive energy network that uses predictive analytics to match energy generation with demand and stores or wastes any extra energy. Additionally, smart meters may collect information about a building's energy consumption and relay that data back to utilities to help load balancing and reduce waste [22]. This minimizes the number of fossil fuels needed to generate energy and reduces the utilities' carbon footprint.

C. *Climate change reduction*

New technologies and applications assist decrease climate change in various domains, including:

i. Energy storage and building automation: IoT-enabled devices can be configured, monitored, and controlled by an intelligent energy storage system to work and consume energy only when necessary. These devices can include lightbulbs, thermostats, and other home applications, as well as heating and cooling systems to decrease expenses and waste, and smart buildings can also automatically modify temperature settings in response to weather changes, dim or turn off lights when no one is in the room, and notify building engineers instantly when a maintenance issue arises [22].

   Using a sensor grid, power consumption could be studied based on previous data to predict day-to-day consumption. This storage layer accumulates the collected data that uses IoT, a machine-learning predictive model that analyses the data, and a service layer that interfaces between the generated model and the building management system. Retrofits improve energy savings in commercial and residential buildings [23].

ii. Traffic: With the help of IoT apps, drivers may find parking spaces more quickly and avoid busy roads, thereby reducing the amount of time they spend on the road and the quantity of carbon dioxide they emit. In addition, a recent Los Angeles project used IoT technology to coordinate traffic lights to improve traffic flow and reduce greenhouse gas emissions, saving over 35 million gallons of fuel yearly [24, 25].

iii. Waste management: When trash cans are full, IoT-connected trash cans are activated in alter collectors so that garbage collection can be made more efficient and thus minimize carbon emissions [26].

iv. Green IoT technology: The majority of current IoT solutions rely on cloud computing. Thousands of IoT devices and equipment need to be connected in most IoT applications, making it difficult to coordinate. Furthermore, the IoT's centralized and server-client structure means that all connected devices can be hacked and exploited, resulting in security risks for the system and privacy concerns for users. In [27], the researchers believe Blockchain could be the answer to this problem. Blockchain's consensus framework requires IoT nodes to confirm they share a goal. Verified transactions are likewise saved in a block connected to the preceding one so information cannot be removed. Everyone can access the transaction history of every node. Any blockchain member immediately knows of any block modifications. Blockchain's distributed ledger can synchronize hundreds of IoT devices. Blockchain consensus techniques provide a secure distributed database. Blockchain can enable decentralized, private-by-design IoT [27]. Despite the use of block technology, it achieves better performance and security, but the number of devices is still large. The growth of the Internet of Things devices will lead to a massive increase in the amount of electrical waste products [28]. Fortunately, the green Internet of Things was born out of these demands (G-IoT). G-IoT is characterized by its energy-efficient characteristics at every stage of its life cycle. Many IoT technologies can benefit from the G-IoT cycle. RFID tags, for example, use radio frequency identification technology. The size of RFID tags has been reduced to reduce the difficult-to-recycle material in each tag. Additionally, green M2M communications are another example of how algorithms and distributed computing technologies can reduce power consumption and improve communication protocols. Sensors in wireless sensor networks can sleep and activate only when necessary. Modulation optimization and cooperative communication techniques can also reduce the node's power consumption. Cluster topologies and multi-path routing are examples of energy-efficient routing systems [27].

3 Dataset Description

IoT pollution monitoring sensor network collects several environmental factors and performs real-time local ozone inference directly on the IoT nodes in the dataset provided by the AIRU POLLUTION MONITORING NETWORK of the University of UTAH [29]. Table 1 shows a dataset sample, which contains about 522,000 readings for IoT nodes. These nodes are sensors for reading the air pollution, humidity, temperature, etc. it also contains the time of reading and the value of Ozone. Table 2 shows the statistics of the dataset.

4 The Proposed Internet of Things Model-Based Machine Learning Techniques for Ozone Prediction

As seen in Fig. 1, the proposed model consists of several steps; reading dataset, data preparation and preprocessing training and testing phases, and the evaluation step.

Table 1 Sample of IoT used dataset

Item	DeviceID	PM1	PM2_5	PM10	MicsRED	MicsNOX	MicsHeater	Temperature	Humidity	Ozone
2019-07-26 22:00:00 + 00:00	209148E036FB	0.091574	0.395162	0.483485	1123.794	1666.015	1	38.94838	29.43662	60
2019-07-26 22:00:00 + 00:00	9884E320942F	0.240911	0.462139	0.537089	1292.886	1571.101	0	38.4881	27.49886	60
2019-07-26 22:00:00 + 00:00	F45EAB08F7A5	1.378342	2.135507	2.392712	1257.616	994.9726	0	41.25575	24.90014	60
2019-07-26 22:00:00 + 00:00	F45EAB9F7485	0.7297	3.386113	3.549413	834.175	933.075	0	38.893	25.63925	60
2019-07-26 22:00:00 + 00:00	606405C95951	0.009701	0.056039	0.108026	1210.948	1901.792	0.025974	39.50714	24.29844	60
2019-07-26 22:00:00 + 00:00	F45EAB9C02DC	0.917923	1.241397	1.424436	1258.167	1637.115	0	40.26359	40.14731	60
2019-07-26 22:00:00 + 00:00	F45EABA254AA	0.465063	0.939896	1.03025	1062.167	853.1458	0	39.80917	26.4075	60
2019-07-26 22:00:00 + 00:00	F45EAB9C466C	0.728225	0.988324	1.135986	1201.38	1044.31	0.028169	41.40563	25.51775	60
2019-07-26 22:00:00 + 00:00	F45EABA27BB6	0.382797	0.788041	1.06477	1101.946	924.5946	0	38.44527	30.64216	60
2019-07-26 22:00:00 + 00:00	F45EAB95EA05	0.059589	0.31337	0.36689	1314.877	1122.425	0	41.96068	27.40603	60
2019-07-26 22:00:00 + 00:00	606405AA0C73	1.310901	1.990753	2.257074	973.8272	1355.481	0	37.51457	29.57333	60
2019-07-26 22:00:00 + 00:00	606405AA0DDE	0.024792	0.251636	0.342558	1075.649	1361.779	0	42.91831	55.88584	60

Table 2 The statistics of the dataset

Item	PM1	PM2_5	PM10	MicsRED	MicsNOX	MicsHeater	Temperature	Humidity	Ozone
Count	416,995	416,995	416,995	416,995	416,995	416,995	416,995	416,995	372,251
Mean	2.57131695	4.045693382	4.60740817	1066.88347	955.819116	0.05726	23.1692	37.4236	27.5696
Std.	5.53603006	8.407867203	9.4428684	184.52017	419.967815	0.18668	10.5605	18.0325	15.9687
Min	0	0	0	337.533333	0	0	− 3.3929	1.51639	0
Max	91.5312	141.4188571	165.024571	4095	2990.8	1	70.9329	99.9907	89

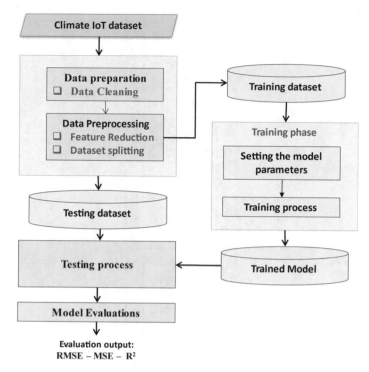

Fig. 1 The structure of the proposed ozone prediction model

4.1 Data Preparing and Preprocessing

a. **Data Preparation**

Table 2 shows that there are 522,000 records in total, but there are different numbers of records for PM1, PM2-5, PM10, MicsRED, MicsNOX, MicsHeater, temperature, humidity, and Ozone. This is a sign that there are errors and missing data. Therefore, we use data cleaning to eliminate irrelevant data, correct structural issues, and handle missing data. Table 3 presents the statistics of the dataset after cleaning the data.

b. **Data preprocessing**

In this stage, a multivariate approach called principal component analysis (PCA) examines a data table in which several interrelated quantitative dependent variables characterize observations. To illustrate the pattern of similarity between the observations and the variables as points in spot maps, it extracts significant information from the statistical data. It represents it as a set of new orthogonal variables called principal components. PCA is mathematically dependent on the singular value decomposition (SVD) of rectangular matrices and the eigen-decomposition of positive semi-definite matrices. The eigenvalues and eigenvectors decide it. Square matrices have vectors

Table 3 The statistics of the cleaned dataset

Item	PM1	PM2_5	PM10	MicsRED	MicsNOX	MicsHeater	Temperature	Humidity	Ozone
Count	285,299	285,299	285,299	285,299	285,299	285,299	285,299	285,299	285,299
Mean	2.810407024	4.403346442	4.987569652	1106.381841	974.1228014	0.057184688	23.16558536	38.28281517	27.82205273
Std.	5.695333347	8.619617963	9.636889041	176.1383585	405.5140083	0.187533411	10.5257356	18.73122642	15.96042141
Min	0	0	0	463.0394737	19.64285714	0	− 2.240833333	1.870779221	0
Max	91.5312	141.4188571	165.0245714	4095	2990.8	1	69.9945614	99.99066667	89

and numbers called eigenvectors and eigenvalues. Together, they offer the eigen-decomposition of a matrix, which examines its structure, including any cross-product, correlation, or covariance matrices [30]. The goals of PCA are to extract the most significant information from the data table; compress the data set by maintaining only this relevant information; Simplify data set description; analyze observations and variables, and reduce the number of dimensions without losing data [30]. So, we'll use the PCA to extract the most important information and reduce the data's dimensionality.

After the data reduction, the dataset was split into two parts for training and testing with ratios of 70% and 30%, respectively.

4.2 Training and Testing Phase

The training and testing processes are done using an ML model called a Decision tree. The decision tree solves typical classification problems correctly. Unlike other nodes, the decision tree classifier's roots contain no incoming edges. "Internal" or "test" nodes have outward edges. Leaves are surviving nodes. Each decision tree internal node splits the instance space using discrete input values. Most tests evaluate one attribute. The attribute's value divides the instance space. Numeric qualities require a decision. Each leaf is assigned using the essential goal value. The leaf may include a probability vector showing the goal value's likelihood. Instances are categorized by the tests undertaken from the tree's root to its leaf [31].

K-fold cross-validation detects overfitting and evaluates model consistency. 5-fold cross-validation is used. Validation partitions data into K equal sets. Each of the K remaining sets is tested once. The proposed prediction model was evaluated to judge the trained model.

4.3 The Evaluation Measures

To evaluate the experimental results, three well-known measures metrics are used, named Mean Squared Error (MSE), the root mean squared error (RMSE), and R-Square (R^2).

MSE is one of the most commonly used measures for regression problems. It's just a calculation of the square difference between the goal and the predicted values of the regression model as in Eq. 1. It penalizes such a slight mistake when it squares the discrepancies, overestimating how poor the model is [32].

$$MSE = \left(\frac{1}{N}\right) \sum_{i=1}^{N} (\hat{y}_i - y_i)^2 \qquad (1)$$

$$RMSE = \sqrt{\left(\frac{1}{N}\right) \sum_{i=1}^{N} (\hat{y}_i - y_i)^2} \tag{2}$$

where N is the number of test samples, y_i is the ith test sample, and $\hat{y}_i$ is the prediction value of y_i.

RMSE is sensitive to aberrant points because it uses the average error. It is used to calculate the difference between the observed and true values as in Eq. 2. If a point's regression value is not plausible, its error is relatively substantial, significantly impacting the RMSE number. The smaller the RMSE, the more accurate the forecast findings will be in general [32].

The fundamental goal of R^2 is to determine how well forecasted and measured data correlate. A dataset contains n values labelled y_1, y_2, ... n, (often designated as y_i or as a vector y = [y_1, y_2, ... n]T), each of which corresponds to a predicted value f1, ..., fn. To calculate the overall amount of squares and the sum of squares remaining Use Eqs. (3) and (4) as follows: Sum total of squares:

$$S_t = \sum_{i=1}^{n} (y_i - \hat{y}_i)^2 \tag{3}$$

Residual sum of squares is sometimes known as the sum of residual squares:

$$S_{re} = \sum_{i=1}^{n} (y_i - f_i)^2 \tag{4}$$

The most common expression of the coefficient of determination is given in Eq. (5).

$$R^2 = 1 - \frac{S_t}{S_{re}} \tag{5}$$

The low values of MSE and RMSE indicate the best result. In contrast, the high value of R^2 indicated a high accuracy [32].

5 Experiments, Results, and Discussion

This section examines the efficacy of the proposed model. The experiments were done on a PC with a Core i7 processor and the MATLAB 2020a software package.

An analysis of the proposed model for making predictions was evaluated to get the definitive verdict of the trained model. The evaluation procedure for the model is carried out both before and after the application of PCA.

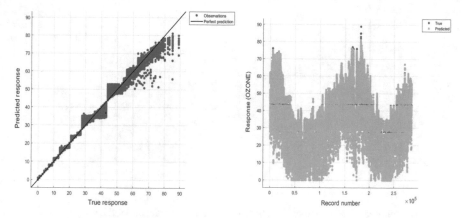

Fig. 2 The training of ozone prediction model without feature reduction

5.1 Experiment I

The decision tree without feature reduction.

The Decision tree is run in the first experiment without feature reduction. The Decision tree with minimum leaf size = 4. Figure 2 shows the training performance of the model. The testing results are 0.89526, 1.00, and 0.71447 for RMSE, R^2, and MSE, respectively.

5.2 Experiment II

Decision tree (with minimum leaf size = 4).

The second experiment uses Decision tree (with minimum leaf size = 4) that used PCA for feature reduction. Figure 3 shows the training performance of the model. The testing results are 0.02139, 1.0, and 0.00045754 for RMSE, R^2 and MSE respectively.

Using PCA for feature reduction improves the prediction model's performance, as the MSE becomes 0.00045754 after 0.71447, and the RMSE value becomes 0.02139 after 0.89526, while R^2 remains 1.0 without change.

6 Conclusion

Climate change is caused by human mismanagement of natural resources, which causes global warming and has a direct and indirect detrimental impact on sustainable development and the surrounding environment. Climate change and global warming can only be prevented if people adjust their lifestyles and use cutting-edge technology

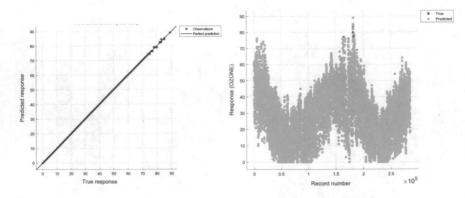

Fig. 3 The training of the ozone prediction model with PCA for feature reduction

like the Internet of Things. Climate change can be combated by predicting specific natural phenomena after collecting appropriate data. The ozone ratio was stud by using an IoT database to track pollution, temperature, and humidity during a certain time perandl as the impact these elements had on ozone levels. In the end, a machinc learning-based model was constructed to forecast ozone levels based on data collected by IoT sensors in the surrounding air. The feature's dimensionality was reduced using PCA in the suggested model. The model's performance is assessed using RMSE, MSE, and R^2. Using PCA to reduce the features improves the prediction model's performance, as the MSE and RMSE drop to 0.00045754 and 0.02139, respectively, after 0.71447 and 0.89526, but R^2 remains unchanged at 1.0.

References

1. https://education.nationalgeographic.org/resource/climate-change
2. Perera, F. P. (2017). Pollution from fossil-fuel combustion is the leading environmental threat to global pediatric health and equity: Solutions exist. *International Journal of Environmental Research and Public Health, 15*(1),16 (2017). https://doi.org/10.3390/ijerph15010016
3. https://climate.nasa.gov/causes/
4. https://www.who.int/health-topics/climate-change#tab=tab_1. Last accessed August 2022.
5. https://www.un-spider.org/news-and-events/news/flooding-affected-more-people-2018-any-other-disaster-type-report-shows
6. Sahya, M., Hasimuna, O. J., Haambiya, L. H., Monde, C., Musuka, C.G., Makorwa, T. H., Munganga, B.P., Phiri, K. J., & Nsekanabo, J. D. (2021). Climate change effects on aquaculture production: Sustainability implications, mitigation, and adaptations. *Frontiers in Sustainable Food Systems, 5* (2021). ISSN: 2571-581X. https://doi.org/10.3389/fsufs.2021.609097
7. Rocque, R. J., Beaudoin, C., Ndjaboue, R., et al. (2021). Health effects of climate change: An overview of systematic reviews. *British Medical Journal Open, 11*, e046333. https://doi.org/10.1136/bmjopen-2020-046333
8. Fawzy, S., Osman, A. I., Doran, J., et al. (2020). Strategies for mitigation of climate change: A review. *Environmental Chemistry Letters, 18*, 2069–2094. https://doi.org/10.1007/s10311-020-01059-w

9. https://dailynewsegypt.com/2022/02/24/egypt-leading-smart-city-adoption-in-africa/. Accessed August 2022.
10. Nižetić, S., Šolić, P., González-de-Artaza D. L., Patrono L. (2020). Internet of Things (IoT): Opportunities, issues and challenges towards a smart and sustainable future. *Journal of Cleaner Production, 274*, 122877 (2020). https://doi.org/10.1016/j.jclepro.2020.122877
11. Zantalis, F., Koulouras, G., Karabetsos, S., & Kandris, D. (2019). A Review of machine learning and IoT in smart transportation. *Future Internet, 11*, 94. https://doi.org/10.3390/fi11040094
12. Shrouf, F., Ordieres, J., & Miragliotta, G. (2014). Smart factories in Industry 4.0: A review of the concept and of energy management approached in production based on the Internet of Things paradigm. In *Proceedings of the 2014 IEEE international conference on industrial engineering and engineering management (IEEM)*, Selangor Darul Ehsan, Malaysia, 9–12 December 2014, pp. 697–701.
13. Bandyopadhyay, D., & Sen, J. (2011). Internet of Things: Applications and challenges in technology and standardization. *Wireless Personal Communications, 58*, 49–69.
14. Atzori, L., Iera, A., & Morabito, G. (2010). The Internet of Things: A survey. *Computer Networks, 54*, 2787–2805.
15. Motlagh, N. H., Bagaa, M., & Taleb, T. (2019). Energy and delay aware task assignment mechanism for UAV-based IoT platform. *IEEE Internet of Things Journal, 6*, 6523–6536.
16. Abd El-Mawla, N., Badawy, M., & Arafat, H. (2019). IoT for the failure of climate-change mitigation and adaptation and IIoT as a future solution world. *Journal of Environmental Engineering, 6*(1), 7–16. https://doi.org/10.12691/wjee-6-1-2
17. Malmodin, J., & Bergmark, P. (2015). Exploring the effect of ICT solutions on GHG emissions in 2030. In *The international conference on informatics for environmental protection (EnviroInfo 2015) third international conference on ICT for sustainability (ICT4S 2015)*.
18. Saeed, N., Al-Naffouri, T. Y., & Alouini, M. (2020). Climate monitoring using internet of X-things, IEEE Internet of Things Magazine, June 2020.
19. https://energypost.eu/internet-things-can-fight-climate-change/. Accessed August 2022.
20. Shafi, U., Mumtaz, R., García-Nieto, J., Hassan, S. A., Zaidi, S. A. R., & Iqbal, N. (2019). Precision agriculture techniques and practices: From considerations to applications. *Sensors, 19*(17), 3796. https://doi.org/10.3390/s19173796.PMID:31480709;PMCID:PMC6749385
21. Lakhiar, I. A., Jianmin, G., Syed, T. N., Chandio, F. A., Buttar, N. I., & Qureshi, W. A. (2018). Monitoring and control systems in agriculture using intelligent sensor techniques: A review of the aeroponic system. *Journal of Sensors, 2018*, 18. https://doi.org/10.1155/2018/8672769.
22. Motlagh, N. H., Mohammadrezaei, M., Hunt, J., & Zakeri, B. (2020). Internet of Things (IoT) and the energy sector. *Energies, 13*, 494. https://doi.org/10.3390/en13020494
23. Ibaseta, D., García, A., Álvarez, M., Garzón, B., Díez, F., Coca, P., Pero, C. D., & Molleda, J. (2021). Monitoring and control of energy consumption in buildings using WoT: A novel approach for smart retrofit. *Sustainable Cities and Society, 65*, 102637. ISSN 2210-6707. https://doi.org/10.1016/j.scs.2020.102637
24. https://ladot.lacity.org/projects
25. https://www.roadsbridges.com/traffic-signals/article/10654415/better-traffic-signals-save-time-and-can-help-save-the-planet
26. Gade, S., & Aithal, S. (2021). Smart city waste management through ICT and IoT driven solution. *International Journal of Applied Engineering and Management Letters (IJAEML), 5*(1). ISSN: 2581-7000. https://doi.org/10.5281/zenodo.4739109
27. Al-Turjman, F., Kamal, A., Rehmani, M. H., Radwan, A., & Pathan, A. K. (2019). The green Internet of Things (G-IoT). *Wireless Communications and Mobile Computing, 2019*. https://doi.org/10.1155/2019/6059343
28. Zhu, C., Leung, V. C. M., Shu, L., & Ngai, E. C. (2015). Green Internet of Things for smart world. *IEEE Access, 3*, 2151–2162.
29. Becnel, T., Kelly, K., Gaillardon, P.-E. (2022) University of Utah AirU pollution monitoring network—Salt Lake City UT—2019-07-26–2021-05-14. IEEE Dataport. https://doi.org/10.21227/aeh2-a413

30. Mishra, S., Sarkar, U., Taraphder, S., Datta, S., Swain, D., Saikhom, R., et al. (2017). Multi-variate statistical data analysis- principal component analysis (PCA). *International Journal of Livestock Research, 7*(5), 60–78. https://doi.org/10.5455/ijlr.20170415115235

31. Abouelmagd, L. M. (2022). E-nose-based optimized ensemble learning for meat quality clas-sification. *Journal of System and Management Sciences, 12*(1), 308–322. https://doi.org/10.33168/JSMS.2022.0122

32. Shams, M. Y., Elzeki. O. M., Abouelmagd, L. M., Hassanien, A. E., Elfattah M. A., & Salem, H. (2021). HANA: A healthy artificial nutrition analysis model during COVID-19 pandemic. *Computers in Biology and Medicine*, S0010-4825(21)00400-5. https://doi.org/10.1016/j.com pbiomed.2021.104606

Emerging Climate Change Technology in Agriculture Sector

Optimized Multi-Kernel Predictive Model for the Crop Prediction with Climate Factors and Soil Properties

Sara Abdelghafar, Ashraf Darwish, and Aboul Ella Hassanien

1 Introduction

The production of crops is essential for the food supply that is expected to be significantly impacted by climate change. For example, decreasing agricultural output may be a result of expected temperature rises, modifications to precipitation patterns, modifications to extreme weather events, and decreases in water availability. For example, depending on the crop's ideal temperature for growth and reproduction, an increase in temperature will have different effects on different crops. The types of crops that are traditionally cultivated there may benefit from warming in some areas, or farmers may be able to switch to crops that are now grown in warmer regions. Conversely, production will decrease if the higher temperature exceeds the crop's optimal temperature [1, 2].

Therefore climate change and food security concerns have pushed the agricultural industry to seek out more innovative ways to increase crop yield productivity. As these difficulties have brought attention to the agricultural landscape's vulnerability, as well as worries about satisfying global food demand sustainably in the face of adversity. The objective is to improve efficiency, which means producing more with fewer resources, which is more important now than ever. The agriculture industry will be transformed by artificial intelligence and machine learning techniques. Farmers

S. Abdelghafar (✉)
School of Computer Science, Canadian International College (CIC), Cairo, Egypt
e-mail: sara_abdelghafar@cic-cairo.com

A. Darwish
Faculty of Science, Helwan University, Cairo, Egypt

A. E. Hassanien
Faculty of Computers and Artificial Intelligence, Cairo University, Cairo, Egypt

S. Abdelghafar · A. Darwish · A. E. Hassanien
Scientific Research Group in Egypt (SRGE), Cairo, Egypt

© The Author(s), under exclusive license to Springer Nature Switzerland AG 2023
A. E. Hassanien and A. Darwish (eds.), *The Power of Data: Driving Climate Change with Data Science and Artificial Intelligence Innovations*, Studies in Big Data 118,
https://doi.org/10.1007/978-3-031-22456-0_10

will have access to the tools and resources they require to maximize the value of each acre.

This study proposes the crop yield prediction model based on the climate change signs, which may be helpful for decisions on the crops that will be grown in agricultural areas based on climatic changes. This might be helpful, for instance, in encouraging the production of new crops in some regions while discouraging the cultivation of others. Predictive models that link crop yields to climate change are crucial for estimating impacts on agriculture as well as the influence of climate change on associated economic and environmental outcomes, which in turn helps inform mitigation and adaptation strategies [3–5].

Support Vector Machine (SVM) is a supervised machine learning technique invented by Boser in 1992 [6], and it can be applied for both classification and regression analysis for high-dimensional datasets with excellent results [7]. SVM is a discriminative classifier algorithm that attempts to find the best separating hyperplane between classes to solve the classification problem [8]. Earlier research used a random selection of SVM Kernel function and hyper-parameter values, which is time-consuming and diminishes the probability of discovering the best option [9]. As a result, the kernel function and hyperparameters that are chosen have a major impact on the sparsity and generalization performance of SVM. Numerous studies have proven the improved SVM for complicated classification and prediction issues like [10–13]. In this work, the proposed optimized predictive model uses Bayesian Optimization (BO) to find the best SVM kernel function and hyper-parameters. The obtained results show that the proposed model BO-SVM performs well in kernel selection and calculating the SVM's optimal hyperparameter values. As a result, BO-SVM predicts agricultural yields with high accuracy using climate change factors.

Following are the key contributions of this study: (1) To effectively tackle climate change and ensure future food security, a decision support tool is proposed to assist farmers and decision-makers in estimating crop yields production based on the climate patterns in their locations; (2) we proposed an optimized crop yield prediction model (BO-SVM) based on the Bayesian optimization utilized for auto kernel selection and optimal parameter estimation of SVM; and (3) Compared to the related work, to the best of our knowledge, this study is the first work to propose a predictive model based on the optimization process for the prediction of crop yield concerned to climate factors and soil properties.

The rest of the paper is organized as follows: Sect. 2 presents some related work in the crop yield predictions based on machine learning techniques. Section 3 introduces the proposed optimized crop yield prediction model. Section 4 defines the experimental dataset and the evaluation metrics and then discusses the obtained experimental results. Finally, Sect. 5 concludes this work and presents some suggestions for future work.

2 Related Work

Earlier, farmers used to choose their crops based on their practical experience. Crop yields are already significantly impacted negatively by climate change. As a result, farmers are unable to choose the best crops due to shifting soil, geographic, and climatic circumstances. Hence, manually estimating the optimum crop to choose for a specific region was more often than not ineffective. As a result, machine learning-based crop production prediction continues to draw a lot of interest from researchers all around the world [14].

Lontsi et al. [15] proposed three frameworks based on three different machine learning techniques; k-Nearest Neighbor, multivariate logistic regression, and decision tree. The three approaches were applied to predict the annual crop yield in some West African nations. The used dataset was collected from different sources, including the food and agriculture organization (FAO) for the UN on the weather, agriculture, pesticides, and chemicals, and the World Bank group's climate change knowledge portal (CCKP). Abbas et al. [16]. Proposes prediction study of potato tuber yield, with the use of four machine learning algorithms— k-nearest neighbor, linear regression, support vector regression (SVR), and elastic net, they forecast potato tuber production from data on soil and crop parameters obtained from proximate sensing. Six fields in Atlantic Canada, including three in Prince Edward Island (PE) and three in New Brunswick (NB), were sampled for soil electrical conductivity, soil moisture content, soil slope, normalized-difference vegetative index (NDVI), and soil chemistry throughout two growing seasons, one in 2017 and the other in 2018. Their findings demonstrate that SVR outperforms all the rest of the models.

Paudel et al. [17] coupled machine learning with agronomic concepts of crop modeling to provide a machine learning baseline for large-scale crop production prediction. The MARS Crop Yield Forecasting System (MCYFS) database was used to construct its features utilizing crop simulation outputs as well as weather, remote sensing, and soil data. Gradient boosting, SVR, and k-nearest neighbors were employed in their proposal to predict the regional yields of the potato, sugar beet, sunflower, soft wheat, spring barley, and other crops in the France, Germany, and the Netherlands.

Another crop modeling study was proposed by Shahhosseini et al. [18], they suggested investigating to demonstrate how to crop modeling and machine learning can be combined to improve estimates of crop production in the US Corn Belt. Their main objectives are to determine whether crop modeling and machine learning can be combined to produce better predictions. They also want to discover which crop modelling parameters can be integrated with machine learning to predict corn production and which hybrid model combinations offer the most accurate predictions. They discovered that using weather information alone is insufficient and that adding simulation crop model characteristics as input parameters to machine learning techniques can improve yield prediction. They indicated that for better yield projections, their suggested machine learning models require additional hydrological inputs.

Another work on the US Corn yield prediction was conducted by Sun et al. [19], they introduced a crop yield prediction model based on a multilevel deep learning model coupling convolutional neural network (CNN) and recurrent neural network (RNN). Their primary objectives were to assess the effectiveness of the suggested approach for predicting Corn Belt yields in the US Corn Belt and to assess the impact of various data sets on the prediction task. Both time-series remote sensing data and data on soil properties were used as inputs. They conducted their experiments in the US Corn Belt states to forecast county-level corn yield from 2013 to 2016. Another work is proposed based on RNN in [20], a hybrid deep learning approach based on an RNN model with long short-term memory (LSTM) was proposed to estimate wheat crop yield in the northern area of India using a 43-year benchmark dataset.

3   The Proposed Optimized Crop Prediction Model

SVM was originally introduced to solve the classification problem of linearly separable data by separating the hyperplane of the distinct classes [21]. Assume we have a training set is open is $\{(x_1, y_1), (x_2, y_2), \ldots, (x_N, y_N)\}$, where x_i represents *ith* sample of the training sample X, and $y_1, y_2, \ldots, y_N$ represent the class labels for $x_1, x_2, \ldots, x_N$.. The primary idea of SVM is to determine the values of the weight vector (w) and the threshold (b) of the decision boundary line that is represented by $w^T x + b = 0$, to make the hyperplane to be as far away from the closest samples and to generate the two planes, H_1 and H_2, as follows:

$$\begin{aligned} H_1 &\rightarrow w^T x_i + b = +1 \ for \ y_i = +1 \\ H_2 &\rightarrow w^T x_i + b = -1 \ for \ y_i = -1 \end{aligned} \tag{1}$$

These two equations can be combined as follows:

$$y_i (w^T x_i + b) - 1 \geq 0 \forall i = 1, 2, \ldots, N \tag{2}$$

where $w^T x_i + b \geq 1$ is the plane for the class ω_+ and $w^T x_i + b \leq 1$ represents the plane for class ω_-, since the hyperplane divides the space into two spaces one is positive and the other is negative, where the classes ω_+ and ω_- are found. The distance from the hyperplane to H_1 and H_2 is given by d_1 and d_2, respectively, and the margin can be determined using $d_1 + d_2 = \frac{2}{\|w\|}$ because the hyperplane is equidistance between the two planes. The goal of the SVM classifier is to maximize the margin width according to Eq. 2 [22]:

$$\begin{aligned} &\min \frac{1}{2} \|w\|^2 \\ &s.t. y_i (w^T x_i + b) - 1 \geq 0 \forall i = 1, 2, \ldots, N \end{aligned} \tag{3}$$

If the data is nonlinearly separable, SVM employs kernel functions to transfer it to a higher dimensional space where it can be linearly separated using a nonlinear function ϕ. The following is how a kernel function is defined [23]:

$$K\left(x_i, x_j\right) = \phi(x_i)^T \phi(x_j) \tag{4}$$

As a result, SVM's objective function will be as follows:

$$\min \tfrac{1}{2}\|w\|^2 + C \sum_{i=1}^{N} \varepsilon_i \tag{5}$$
$$s.t. y_i\left(w^T \phi(x_i) + b\right) - 1 + \varepsilon_i \geq 0 \forall i = 1, 2, \ldots, N$$

where ε_i is a slack variable added to the objective function to relax the linearity constraints, and each ε_i represents the distance between the *ith* training sample and the corresponding margin hyperplane; when $0 \leq \varepsilon_i \leq 1$, the sample is correctly classified because it is between the margin and the correct side of the hyperplane; when $\varepsilon_i > 1$, the sample is misclassified. [24].

Bayesian optimization is a class of machine-learning-based sequential optimization methods of uncertain objective functions, and can be represented as follows [25],

$$x^* = \underset{x \in X}{\operatorname{argmax}} f(x) \quad where\ X \subseteq R^D \tag{6}$$

where the objective is to find the parameters x^* that maximizes the function $f(x)$ over some domain X consisting of finite lower and upper bounds on every variable.

BO is a sequential search methodology that combines both exploration and exploitation and is typically much more effective than grid search or random search. The search framework is composed of two main parts: a Bayesian statistical model that models the objective function and an acquisition function that determines where to sample next. The Bayesian posterior probability distribution provided by the statistical model, which is typically a Gaussian process, describes possible values for $f(x)$ at a candidate point x. Then posterior distribution is updated each time f is observed at a new location, the acquisition function calculates the value that would result from evaluating the objective function at a new point x. The acquisition function is always balancing two important aspects: investigating regions with significant epistemic uncertainty about the function and leveraging regions with high predictive mean [26, 27].

Suppose that the hyper parameter optimization function $f(x)$ follows the Gaussian process, then is $p(f(x)|x, H)$ a normal distribution. Based on the results of existing N group experiments, $H = \{x_n, y_n\}_{n=1}^{N}$, BO is modeled as a Gaussian process, and the posterior distribution $p(f(x)|x, H)$ of $f(x)$ is calculated [28].

After getting the posterior distribution of the objective function, an acquisition function $a(x, H)$ is created to compromise between sampling at places where the model predicts a high objective and sampling at regions where the prediction uncertainty is high. The objective of choosing the next sampling point is to maximize the acquisition function. Suppose $y^* = \min y_n$, $1 \leq n \leq N$ is the optimal value in the currently existing sample, the desired improvement function is as follows [29],

$$a(x, H) = \int_{-\infty}^{\infty} \max(y^* - y, 0) P(y|x, H) dy \tag{7}$$

In the proposed model, BO constructs a probability model of the objective function and employs it to choose the optimum kernel and its hyperparameters for evaluating the prediction goal function. The objective function is modeled using a Gaussian process as a probabilistic measure, and BO gives robust optimization solutions. The optimum kernel function is selected during the optimization phase from a variety of kernel functions, including Linear, Gaussian, Quadratic, and Cubic.

The penalty parameter (C) and the parameters of the selected kernel (σ) are the main SVM hyperparameters that significantly affect classification accuracy. C controls how many outliers are considered when calculating Support Vectors (SVs), and as a result, changing values of C control classification accuracy by striking a balance between margin maximization and error minimization. While the kernel parameters, used as tuning parameters to improve classification accuracy, have an impact on the feature space transformation map. As a result, BO's main objectives in the suggested model are to select the kernel function and then figure out the optimal values for C and σ to acquire the lowest testing classification error.

The SVM kernel and hyper parameters were initially initialized in the proposed model with random selections to begin the training process for developing the trained classifier model that will be evaluated using the evaluation phase. Until the termination criteria are met, which is indicated by the determination of the optimal fitness value or the maximum number of iterations, BO searches for the global optimum solution over the iterations. By minimizing the classification error, accuracy is employed as the fitness function to calculate the optimal fitness value. As presented below in Algorithm 1 and Fig. 1.

Algorithm 1: optimized predictive model BO-SVM

Input: $D = (x_i, y_i)$, $i = 1, \ldots, N$: Training dataset with x_i is input y_i is target class label, $f(x) = \frac{TP}{TP+FP}$: objective Function, $a(x, H)$: acquisition function, $Max Iter$: Maximum number of iterations, and $Max Fit$: Best fitness value

---

Output: Y_p: predicted class label and $f(x)$.

(continued)

(continued)

Algorithm 1: optimized predictive model BO-SVM

$H \leftarrow \theta$;
Initialize SVM with random kernel and initialized parameters c and σ;
Train SVM on the training dataset based on random kernel and initialized parameters;
Obtain the predicted target class label Y_p;
Calculate f(x);
Random initialization of Gaussian process, calculate $p(f(x)|x, H)$;
Initialize t $\leftarrow$ 0;

while t < MaxIter and fitness value < MaxFit **Do**

$\dot{x} \leftarrow \arg \max_x a(x, H)$;

Evaluate $\dot{y} = f\left(\dot{x}\right)$;

$H \leftarrow H \cup \left(\dot{x}, \dot{y}\right)$;

Remodeling Gaussian process according to H, calculate $p(f(x)|x, H)$;
Update SVM kernel and parameters c and σ;
Test SVM based on updated kernel parameters;
Obtain Y_p;
Calculate f(x);
$t \leftarrow t + 1$;
end;
Return Y_p and $f(x)$

4 Results, Discussion, and Analysis

This section presents the procedures for implementing the proposed model, the dataset used to evaluate the performance of the proposed optimized model BO-SVM and analyzes the results obtained. The proposed model was implemented using MATLAB version 2022a with statics and the machine learning toolbox. As a preprocessing step, the experiment was conducted after standardization. The dataset was split into 70%, 15%, and 15% for training, validation, and testing respectively. The one-vs-one is selected as a multi-classification method. The maximum number of iterations in the optimization phase is 30.

4.1 Dataset Description

The proposed model's efficacy is proved using the Crop Recommendation Dataset that is published by Kaggle [30]. The dataset size is 2200 records and has a certain number of attributes for some of the selected climate factors and soil properties; the

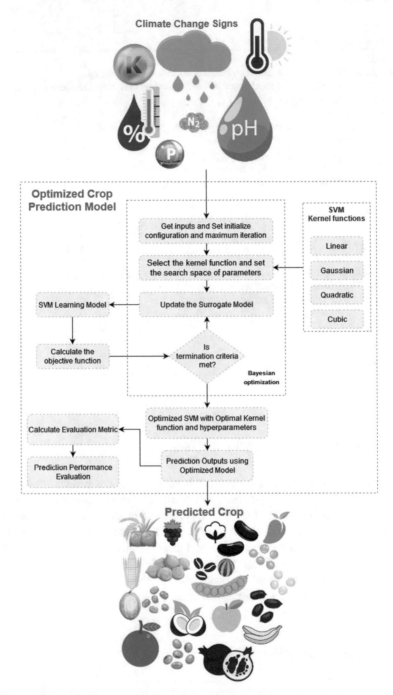

Fig. 1 The proposed optimized crop prediction model BO-SVM flowchart

ratio of Nitrogen content in soil (N), the ratio of Phosphorous content in soil (P), the ratio of Potassium content in soil (K), Temperature (the typical soil temperature for bioactivity ranges from 50 to 75F), Ph (A scale that is used to determine if something is acidic or basic (Acid Nature- Ph < 7; Neutral- Ph = 7; Base Nature-P > 7)), relative Humidity, and Rainfall in mm.

The class label field contains 22 crop types; Apple, Banana, Black gram, Chickpea, Coconut, Coffee, Cotton, Grapes, Jute, Kidney beans, Lentil, Maize, Mango, Moth-beans, Mungbeans, Muskmelon, Orange, Papaya, Pigeonpeas, Pomegranate, Rice, and Watermelon. Samples of the dataset are shown in Table 1.

4.2   Experimental Results and Analysis

The crop type is the response of the proposed predictive model based on the seven attributes of the climate factors and soil properties. The evaluation measures are reported in the form of the confusion matrix which is typically used for demonstrating the performance of a classification model on validation and test sets, in which the correct predictions are determined by mapping predicted outputs to actual outputs, where True Positive (TP) represents all values are correctly predicted and False Negative (FN) represents all values are incorrectly classified. The classification accuracy is measured by the proportion of the number of true predictions out of all predictions made [31, 32]. The obtained results in the form of the confusion matrix are presented in Fig. 2 and 3. The confusion matrix reveals that the proposed model BO-SVM achieved high classification accuracy, as presented in Table 2.

Also, the outputs of the optimization phase are reported by plotting the minimum observed and estimated function values versus the number of function evaluations, as shown in Fig. 4.

5   Conclusion and Future Work

Climate has a significant influence on crop yields. Therefore, accurate assessments of how climate change may affect crop yields are critical for ensuring global food security. In this study, a crop yield prediction model based on signs of climate change is proposed. This model could be useful for selecting the crops that will be produced in agricultural areas in response to climate change. This could be useful, for example, in promoting the growth of new crops in particular places while inhibiting the cultivation of others. Also, predictive models that link crop yields to climate change are essential for estimating the impacts of climate change on related economic and environmental outcomes. The optimized predictive model is proposed based on the optimization framework for the multi-kernel SVM using Bayesian Optimization to obtain the optimal auto kernel and hyper-parameters selection. The obtained results demonstrate that the proposed model BO-SVM is capable of selecting the SVM's

Table 1 Samples of the dataset

N	P	K	Temperature	PH	Humidity	Rainfall	Label
35	134	204	9.95	5.84	82.55	66.01	Grapes
49	69	82	18.32	7.26	15.36	81.79	Chickpea
22	60	24	18.78	5.63	20.25	104.26	Kidneybeans
90	42	43	20.88	6.50	82.00	202.94	Rice
27	63	19	20.93	5.56	21.19	133.19	Kidneybeans
85	58	41	21.77	7.04	80.32	226.66	Rice
71	54	16	22.61	5.75	63.69	87.76	Maize
24	128	196	22.75	5.52	90.69	110.43	Apple
60	55	44	23.00	7.84	82.32	263.96	Rice
136	36	20	23.10	6.93	84.86	71.30	Cotton
15	11	38	23.13	6.63	92.68	109.39	Pomegranate
7	144	197	23.85	6.13	94.35	114.05	Apple
13	5	8	23.85	7.47	90.11	103.92	Orange
33	77	15	23.90	7.80	66.32	40.75	Lentil
22	17	5	24.12	6.95	90.72	102.84	Orange
133	47	24	24.40	7.23	79.20	90.80	Cotton
16	8	9	24.60	7.60	91.28	111.29	Orange
109	21	55	24.90	6.77	89.74	57.45	Watermelon
63	41	45	25.30	7.12	86.89	196.62	Jute
89	47	38	25.52	6.00	72.25	151.89	Jute
37	5	34	25.79	5.78	93.84	152.42	Coconut
61	44	17	26.10	6.93	71.57	102.27	Maize
91	21	26	26.33	7.26	57.36	191.65	Coffee
107	21	26	26.45	7.24	55.32	144.69	Coffee
60	37	39	26.59	6.03	82.94	161.25	Jute
25	62	21	26.73	7.04	68.14	67.15	Blackgram
102	14	52	26.79	6.51	89.65	57.74	Watermelon
19	26	29	26.93	5.67	98.80	166.57	Coconut
82	75	55	27.35	6.28	78.49	92.16	Banana
115	17	55	27.58	6.78	94.12	28.08	Muskmelon
3	49	18	27.91	3.69	64.71	32.68	Mothbeans
25	48	21	28.44	6.27	83.49	52.55	Mungbean
56	79	15	29.48	7.45	63.20	71.89	Blackgram
117	81	53	29.51	5.51	78.21	98.13	Banana

(continued)

Table 1 (continued)

N	P	K	Temperature	PH	Humidity	Rainfall	Label
2	40	27	29.74	5.95	47.55	90.10	Mango
38	61	21	30.27	4.70	67.39	127.78	Pigeonpeas
39	24	31	33.56	4.76	53.73	98.68	Mango
43	64	47	38.59	6.83	91.58	102.27	Papaya
42	59	55	40.10	6.98	94.35	149.12	Papaya

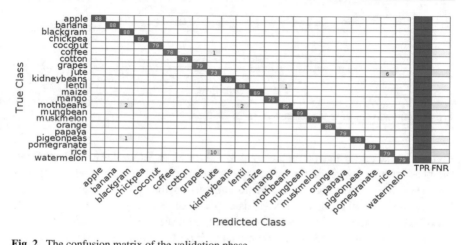

Fig. 2 The confusion matrix of the validation phase

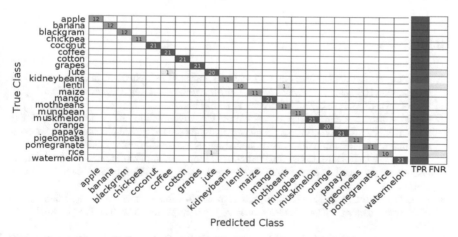

Fig. 3 The confusion matrix of the test phase

Table 2 The overall classification accuracy of the validation and test sets

	Validation	Test
	TP = 1833 FP = 23	TP = 341 FP = 3
Accuracy (%)	98.8	99.1

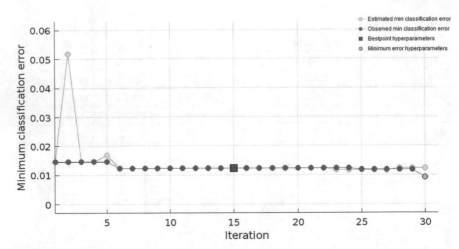

Fig. 4 Minimum objective versus the number of iterations

optimal kernel and hyper-parameter values. Accordingly, BO-SVM provides reliable predicted results for predicting crop type based on climate change and soil feature attributes. In future work, more climate and soil attributes will be added such as; meteorological variations data, wind data, pollution data ...etc. that can probably maximize the value of the expected outcomes from the prediction model. Also, we are interested to apply the prediction model to Egypt and African agriculture cases.

References

1. Zhao, C., et al. (2017). Temperature increase reduces global yields of major crops in four independent estimates. *The Proceedings of the National Academy of Sciences (PNAS), 114*(35), 9326–9331. https://doi.org/10.1073/pnas.1701762114
2. Global Climate Change Impact on Crops Expected Within 10 Years, NASA Study Finds. [Online]. Available: https://climate.nasa.gov/news/3124/global-climate-change-impact-on-crops-expected-within-10-years-nasa-study-finds. Accessed on: July 1, 2022.
3. Yadav, K., & Geli, H. M. E. (2021). Prediction of crop yield for New Mexico based on climate and remote sensing data for the 1920–2019 period. *Land, 10*, 1389. https://doi.org/10.3390/land10121389
4. Leng, G., & Huang, M. (2017). Crop yield response to climate change varies with crop spatial distribution pattern. *Science and Reports, 7*, 1463. https://doi.org/10.1038/s41598-017-01599-2
5. Lobell, D. B., & Asseng, S. (2017). Comparing estimates of climate change impacts from process-based and statistical crop models. *Environmental Research Letters, 12*(2017), 015001. https://doi.org/10.1088/1748-9326/015001
6. Boser, B.E., Guyon, I.M., & Vapnik, V.N. (1992). A training algorithm for optimal margin classifiers. *Proceedings of the fifth annual workshop on Computational learning theory* (pp. 144–152, 1992). Pittsburgh.
7. Hearst, M. A., Dumais, S. T., Osuna, E., Platt, J., & Scholkopf, B. (1998). Support vector machines. *IEEE Intelligent Systems and their Applications, 13*(4), 18–28.

8. Chen, J., & Licheng, J. (2000). Classification mechanism of support vector machines. *Proceedings of the fifth International Conference on Signal Processing Proceedings.* 16th World Computer Congress 2000, IEEE.

9. Vatsa, M., Singh, R., & Noore, A. (2005). Improving biometric recognition accuracy and robustness using dwt and svm watermarking. *IEICE Electronics Express, 2*(12), 362–367. https://doi.org/10.1587/elex.2.362

10. Tharwat, A., Hassanien, A. E., & Elnaghi, B. E. (2017). A BA-based algorithm for parameter optimization of Support Vector Machine. *Pattern Recognition Letters, 93*, 13–22. https://doi.org/10.1016/j.patrec.2016.10.007

11. Tharwat, A., & Hassanien, A. E. (2018). Chaotic antlion algorithm for parameter optimization of support vector machine. *Applied Intelligence, 48*, 670–686. https://doi.org/10.1007/s10489-017-0994-0

12. Abdelghafar, S., Goda, E., Darwish, A., & Hassanien, A.E. (2019). Satellite lithium-ion battery remaining useful life estimation by coyote optimization algorithm. In *Proceedings of 2019 Ninth International Conference on Intelligent Computing and Information Systems (ICICIS)* (pp. 124–129). IEEE. https://doi.org/10.1109/ICICIS46948.2019.9014752.

13. S. Abdelghafar, A. Darwish, A. E. Hassanien, "Cube Satellite Failure Detection and Recovery Using Optimized Support Vector Machine", in Proc. International Conference on Advanced Intelligent Systems and Informatics, Springer, 664–674, 2018. https://doi.org/10.1007/978-3-319-99010-1_61.

14. Suruliandi, A., Mariammal, G., & Raja, S. P. (2021). Crop prediction based on soil and environmental characteristics using feature selection techniques. *Mathematical and Computer Modelling of Dynamical Systems, 27*(1), 117–140. https://doi.org/10.1080/13873954.2021.1882505

15. Cedric, L.S., Adoni, W.Y.H., Aworka, R., Zoueu, J.T., Mutombo, F.K., Krichen, M., & Kimpolo, C.L.M. (2022). Crops yield prediction based on machine learning models: Case of West African countries. *Smart Agricultural Technology 2*. https://doi.org/10.1016/j.atech.2022.100049.

16. Abbas, F., Afzaal, H., Farooque, A.A., Tang, S. (2020). Crop yield prediction through proximal sensing and machine learning algorithms. *Agronomy 10*(7). https://doi.org/10.3390/agronomy10071046.

17. Paudel, D., Boogaard, H., de Wit, A., Janssen, S., Osinga, S., Pylianidis, C., & Athanasiadis, I. N. (2021). Machine learning for large-scale crop yield forecasting. *Agricultural Systems, 187*, 103016. https://doi.org/10.1016/j.agsy.2020.103016

18. Shahhosseini, M., Hu, G., Huber, I., & Archontoulis, S. (2021). Coupling machine learning and crop modeling improves crop yield prediction in the US corn belt. *Science Reports 11*. https://doi.org/10.1038/s41598-020-80820-1.

19. Sun, J., Lai, Z., Di, L., Sun, Z., Tao, J., & Shen, Y. (2020). Multilevel deep learning network for county-level corn yield estimation in the US Corn Belt. *IEEE Journal of Selected Topics in Applied Earth Observations and Remote Sensing 13*, 5048–5060. https://doi.org/10.1109/JSTARS.2020.3019046.

20. Bali, N., & Singla, A. (2021). Deep learning based wheat crop yield prediction model in Punjab region of North India. *Applied Artificial Intelligence, 35*(15), 1304–1328. https://doi.org/10.1080/08839514.2021.1976091

21. Drucker, H., Wu, D., & Vapnik, V. N. (1999). Support vector machines for spam categorization. *IEEE Transactions on Neural Networks, 10*(5), 1048–1054.

22. Wang, L. (2005). Support vector machines: Theory and applications (vol. 177). Springer Science and Business Media.

23. Chapelle, O., Vapnik, V., Bousquet, O., & Mukherjee, S. (2002). Choosing multiple parameters for support vector machines. *Machine Learning, 46*, 131–159.

24. Scholkopf, B., & Smola, A.J. (2001). Learning with Kernels: Support vector machines, regularization, optimization, and beyond. MIT press.

25. Frazierar, P.I. (2018). A tutorial on Bayesian optimization. Xiv:1807.02811. https://doi.org/10.48550/arXiv.1807.02811.

26. Joy, T.T., Rana, S., Gupta, S., & Venkatesh, S. (2020). Fast hyperparameter tuning using Bayesian optimization with directional derivatives. *Knowledge-Based Systems 205*. https://doi.org/10.1016/j.knosys.2020.106247.
27. Shahriari, B., Swersky, K., Wang, Z., Adams, R. P., & de Freitas, N. (2016). Taking the human out of the loop: A review of Bayesian optimization. *Proceedings of the IEEE, 104*(1), 148–175. https://doi.org/10.1109/JPROC.2015.2494218
28. Snoek, J., Larochelle, H., & Adams, R.P. (2012). Practical Bayesian optimization of machine learning algorithms. *Advances in Neural Information Processing Systems*, 2951–2959.
29. Guoc, H., Zhuangc, X., Liangc, D., & Rabczuk, T. (2020). Stochastic groundwater flow analysis in heterogeneous aquifer with modified neural architecture search (NAS) based physics-informed neural networks using transfer learning. *International Journal of Engineering Science.* https://doi.org/10.48550/arXiv.2010.12344.
30. Crop Recommendation Dataset. [Online]. Available: https://www.kaggle.com/datasets/atharvaingle/crop-recommendation-dataset. Accessed on: June 15, 2022.
31. Abdelghafar, S., Darwish, A., Hassanien, A. E., Yahia, M., & Zaghrout, A. (2019). Anomaly detection of satellite telemetry based on optimized extreme learning machine. *Journal of Space Safety Engineering, 6*(4), 291–298. https://doi.org/10.1016/j.jsse.2019.10.005
32. Ezzat, D., Hassanien, A. E., Darwish, A., Yahia, M., Ahmed, A., & Abdelghafar, S. (2021). Multi-objective hybrid artificial intelligence approach for fault diagnosis of aerospace systems. *IEEE Access, 9*, 41717–41730. https://doi.org/10.1109/ACCESS.2021.3064976

An Intelligent Crop Recommendation Model for the Three Strategic Crops in Egypt Based on Climate Change Data

Sally Elghamrawy, Athanasios V. Vasilakos, Ashraf Darwish, and Aboul Ella Hassanien

1 Introduction

As the world's population exponentially grows, the significance of intelligent agriculture increases. The population growth has doubled to more than 7.2 billion people, reports the UN Food and Agriculture Organization [1]. Food security is confronted with extraordinary difficulties by many factors such as climate change, crop diseases, etc.

The demand for global food is predicted to double by 2050 [2]. To meet growing food demand in the presence of global warming need of great side of the climate change factors influencing food production. It is essential to inspect crop yield response to climate inconsistency. Typically, farmers would be adjustable to the continuing changes in climate conditions than risky actions, which requires a deep understanding of the impacts of climate excesses on agricultural production. Based on World Meteorological Organization reports [3, 4], the impacts of climate change

S. Elghamrawy (✉)
Computer Engineering Department, MISR Higher Institute for Engineering and Technology, Mansoura, Egypt
e-mail: Sally_elghamrawy@ieee.org

A. V. Vasilakos
Center for AI Research (CAIR), University of Agder (UiA), Grimstad, Norway
e-mail: thanos.vasilakos@uia.no

A. Darwish
Faculty of Science, Helwan University, Cairo, Egypt
e-mail: ashraf.darwish.eg@ieee.org

A. E. Hassanien
Faculty of Computers and Artificial Intelligence, Cairo University, Cairo, Egypt

S. Elghamrawy · A. Darwish · A. E. Hassanien
Scientific Research Group in Egypt (SRGE), Cairo, Egypt

© The Author(s), under exclusive license to Springer Nature Switzerland AG 2023
A. E. Hassanien and A. Darwish (eds.), *The Power of Data: Driving Climate Change with Data Science and Artificial Intelligence Innovations*, Studies in Big Data 118,
https://doi.org/10.1007/978-3-031-22456-0_11

increased from 2015 to 2021. Machine learning has been used by different researchers in different sectors [5–9] in our life for presenting intelligent models that can predict and recommend the best decisions to help decision-makers.

Maize, wheat, and rice are one of the most significant grownup cereals in Egypt, and there is a large gap between their production and consumption [10]. Currently, several researchers comprehend some artificial intelligence techniques to predict crop yield production based on the current and near-future climate change data.

1.1 Motivation

Climate change is the major factor impacting the agricultural production. The potential variations of climate change attributes such as temperature, humidity, precipitation intensity, and CO_2 concentrations directly impact the crop yield production. Artificial Intelligence (AI) models have a superior prospective to predict, classify and recognize climate change threats.

Development has been made in the field of deep learning in the latest years for mitigating different climate change challenges that affect the crop yield production.

The agriculture production will be enhanced by artificial intelligence and machine learning techniques. The main goal is to increase and enhance the crop production in spite of the climate change effects. Many researchers attempt to predict\recommend the best crop to be planted in a specific climate change scenario by detecting the most climate change's feature that affect the crop production.

1.2 Main Contributions

Thus, the main contributions of this work is as follows:

- An Automated Crop Recommendation Model (ACRM) is proposed to recommend the most suitable crop to be planted based on Climate change data using Convolutional Neural Network (CNN) as a deep learning technique
- Utilizing the deep learning technique in the model bounces the capability to analyze a large number of features and examine diverse data types for the training phase.
- The model targets three strategic crops growing in Egypt (Maize, Wheat, and rice), which will help the decision makers choose a suitable choice based on analyzed data.
- An optimized Conventional Neural Network is implemented in the crop recommendation model for predicting the most influential factors that affect crop yield production using the Grey Wolf Optimization algorithm (GWO). Unlike other pre-trained models, this model is trained from scratch using different layers.

The paper is organized as follows: Sect. 2 shows the background and literature review. The Climate Change Scenarios used are defined in Sect. 3. Section 4 describes the proposed Automated Crop Recommendation Model (ACRM) and the proposed optimized Conventional Neural Network using GWO. The performance of ACRM is estimated in Sect. 5. Section 6 argues the results and concludes the paper.

2 Background and Literature Review

Climate change greatly impacts crop yield production by direct, indirect, and socioeconomic effects, as shown in Fig. 1. Several studies [11–16] presented a prediction and recommendation model for crop yield production to identify the climate change impacts and determine the most significant features affecting crop yield production.

Different machine learning applications were presented in anomaly detection and pattern recognition. Gaitán [11] put a spot on the recent environmental applications that directly impact agricultural yield production, particularly for the weather and climate conditions. Droesch [12] proposed a new approach, namely semiparametric neural networks (SNN), that utilize different parameters of a deep neural network to crop yield modeling, which can be used in high dimensional datasets. Mehnatkesh et al. [13] proposed an optimized artificial neural network model to recognize the most significant topographic features and soil characteristics that impacted the unevenness in nominated wheat yield in Iran [13].

In [14], the authors presented a recommendation system that predicts the best yield for the farmer and recommends pest control techniques. The authors compared the SVM classification, Logistic Regression, and Decision Tree algorithm, and their results indicated that SVM classification gives better accuracy than other algorithms. Ransom et al. [15] presented and evaluated different machine learning algorithms for

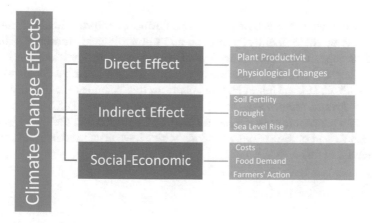

Fig. 1 Related work on climate change effect on agriculture

employing climate change data for refining corn production. The authors evaluated the stepwise, elastic net regression, ridge regression, principal component regression (PCR), least absolute shrinkage and selection operator (Lasso), partial least squares regression (PLSR), decision tree, and random forest algorithms in their research. Vasu et al. [16] presented a nutrient management system for farms. The proposed system's main goals are to evaluate the grade of soil pH and the organic carbon, recognize the insufficiency regions for exact nutrients, and detect the significant features of soil fertility.

Plant Productivity is one of the direct effects of climate change. With the unexpected deviations in environmental circumstances, the severe effects on plant productivity are developing in boundless strengths due to climate change's direct and indirect effects. In addition, soil fertility, air pollution, water accessibility, and physiological changes affect agriculture productivity [17]. It is predicted that, in the near future, the productivity of the main crops is appraised to be decreased in many nations due to global warming and other environmental impacts [18, 19]. The Abscisic Acid (ABA) [20], is the main hormone involved in regulating the responses to numerous abiotic stresses and also triggers many physiological approaches in plants as water shortage and products many reactive genes [21, 22].

The indirect effect of climate change includes fertilizers, which are critical to moderate the impact of global warming. It offers significant energy to plants to sustain the soil's fertility and increase yield [23]. Drought, high temperatures, and sea level are vital factors of the indirect effect of climate change, impacting crop yields [24]. The mutual influence of heat and drought strains on crop yield has been examined and evaluated on different crops, and the results showed that it had more harmful effects than different stress [25, 26].

3   Climate Change Scenarios

A climate scenario is a reasonable representation of future climate, using some assumptions of the amount of radiation and climatological information that can be used as input to climate change models. Distributing knowledge on human-induced climate change [27, 28] is the role of the Intergovernmental Panel on Climate Change—IPCC. It is a United Nations platform that specifies the number of acute effects on several sectors associated with climate change. IPCC approved some climate change scenarios as the Representative Concentration Pathway (RCP) and the Special Report on Emissions Scenarios, which have been used in the experiments implemented in this research.

Table 1 Different representative concentration pathway parameters

RCP	Radiative forcing values (W/m^2)	Average future temperature increase (°C)	Average future sea level rise (m)
RCP2.6	2.6	1.0	0.4
RCP4.5	4.5	1.8	0.47
RCP6	6	2.2	0.48
RCP8.5	8.5	3.7	0.63

Fig. 2 Different RCP and future prediction

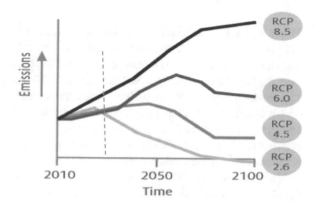

3.1 The Representative Concentration Pathway

Four pathways were utilized to describe different climate futures based on the Greenhouse Gases (GHG) volume emitted in the years to come [29]. The Representative Concentration Pathways (RCPs) imprison these future developments. The RCPs are RCP2.6, RCP4.5, RCP6, and RCP8.5 [30, 31] categorized based on a series of radiative forcing values in the year 2100 (2.6, 4.5, 6, and 8.5 W/m^2, respectively), as shown in Table 1 and Fig. 2. RCP forecasts how concentrations of greenhouse gases in the air will vary in the future due to harmful human activities.

3.2 The Special Report on Emissions Scenarios

The IPCC devolved a report, namely the Special Report on Emissions Scenarios (SRES), which deals with the expectation of radiation trajectories and climate impacts. Unlike the RCPs that repair the radiations trajectory and subsequent radiative imposing relatively than the socioeconomic environments.

The GHG radiation scenarios defined before have been utilized to predict likely upcoming climate change. The SRES scenarios were used in the IPCC Third Assessment Report (TAR), published in 2001, and the IPCC Fourth Assessment Report (AR4), published in 2007.

4 The Proposed Automated Crop Recommendation Model Using Deep Learning

The proposed Automated Crop Recommendation Model (ACRM) is based on Climate change data, using an optimized Convolutional Neural Network (CNN) as a deep learning technique. The main objective of the proposed model is to recommend the most suitable crop to be planted in a specific time based on the climate change data. The model analyzes factors such as CO_2 concentrations, climate change scenarios, crop projected yield, temperature, etc. Our model allows the decision makers to have an overview of the suitability of crop growing in the long and short future. ACRM consists of six main modules: pre-processing, mapping, deep learning, Grey Wolf optimization, recommendation, and evaluation (Fig. 3).

The crop data and the climate change data are combined in one dataset using the mapping module that checks the compatibility of the dataset and stores the data in a proposer format. In the pre-processing module, the missing data in the dataset will be handled by statically replacing it. Any redundancy accrued in the dataset will be removed. The deep learning module presents a proposed optimized CNN model to extract the dataset's significant factors. It consists of 18 layers: convolution

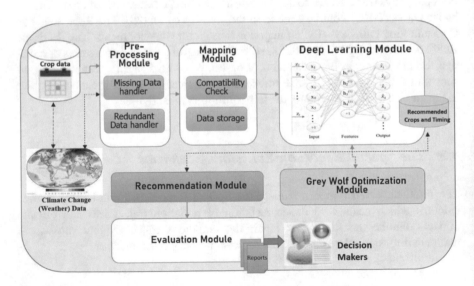

Fig. 3 The proposed automated crop recommendation model using deep learning

Table 2 The details of the proposed CNN architecture model

Name	Type	Size
CONV1	Convolution	3 *3 pixels—16 different filters—128 bias
ELU1	Exponential linear units	74 × 74
POOL1	Max pooling	3 × 3 max pooling layer
CONV2	Convolution	3 × 3 pixels—32 different filters——256 bias 5 × 5 convolutional kernels
ELU2	Exponential linear units	9 × 9
P0012	Max pooling	3 × 3 max pooling layer
CONV3	Convolution	3 pixels—32 different filters—256 bias
ELU3	Exponential linear units	3 × 3
P0OL3	Max pooling	3 × 3 max pooling layer
CONV4	Convolution	3 × 3 pixels—32 different filters——512 bias
ELU4	Exponential linear units	1 × 1
POOL4	Max pooling	3 × 3 max pooling layer
FULCON1	Fully connected	64 neurons
DROPOUT1	Dropout	with a probability of 0.1
FULCON2	Fully connected	48 neurons
DROPOUT2	Dropout	With a probability of 0. 1
FULCON3	Fully connected	3 neurons: 3 crop recommendation
SOFIMA1	Softmax	1 × 1

layers, max-pooling layers, Exponential Linear Units (ELU) layers, fully connected layers, dropout layers, and output layer, Softmax. The details of the proposed CNN architecture model are shown in Table 2.

4.1 The Optimized Conventional Neural Network Using the Grey Wolf Optimization Algorithm (GWO)

The Grey Wolf Optimization (GWO) algorithm is used to optimize the CNN model. The model's weights are updated and optimized using GWO. At the beginning of the implementation, the CNN's hyperparameters are initialized: the initial learning rate (α), the initial parameter vector (w_0) and the maximum iterations (Max_{it}) using Adam suggested parameters' initializations.

The proposed recommendation module's main task is to recommend the crop to be planted by measuring how probable a specific crop is to be planted in a given condition and climate change scenario. The model compares the probability of crops like maize, rice, and wheat and ranks them according to the best choice. The Softmax activation function is used in the output layer. The recommendation degree can be classified into [1. Highly Recommended (HR), 2. Moderate Recommended (MR), 3. Not Recommended (NR)].

In addition, an optimized Conventional Neural Network is implemented in the crop recommendation model for predicting the most influential factors that affect crop yield production, using the Grey Wolf Optimization algorithm (GWO), as shown in Fig. 4.

The CNN model is trained to recommend the relevant climate change factors that impact crop production, and the results obtained from the model are evaluated using the evaluation module.

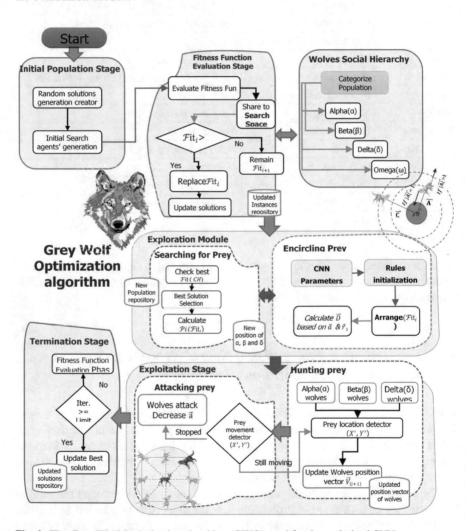

Fig. 4 The Grey Wolf Optimization algorithm (GWO) used for the optimized CNN

5   The Experimental Results

Several experiments are used to evaluate the proposed ACRM. The optimized CNN implemented in the ACRM module is used along with different models over the same dataset for a comparative study to test the performance of ACRM. The results obtained are investigated for the three strategic crops in Egypt (maize, wheat, and rice) over the test data. The recommendation degree obtained is classified into:

1. Highly Recommended (HR)
2. Moderate Recommended(MR)
3. Not Recommended(NR)

5.1   The Dataset Used in the Experiments

The experiments are implemented on the dataset [32] for Egypt data that lies at 36°8' north latitude and 30° 8' east longitude. The data set consists of climate change data such as different Time, adaptation, CO_2, soil, and climate scenarios, along with maize, wheat, and rice crop data, as shown in Table 3.

5.2   Performance Evaluation Measures

Several experiments were used to validate the effectiveness of the proposed ACRM model's performance on an Intel Core i5-8250 1.80 GHz processor with 8 GB memory. Python 3.7 and TensorFlow 1.3 are used to run the model. Table 4 shows the confusion matrix used to evaluate the performance.

$$Overall\,Accuracy = ACC = \frac{TP + TN}{TP + FP + TN + FN}$$

5.3   Experiment 1: The Projected Impacts on the Three Strategic Crops Using Different Climate Scenarios

This experiment measures the projected impacts on the three strategic crops (Wheat, Maize, and Rice) in Egypt using different climate scenarios RCP2.6, RCP4.5, RCP6.0, RCP8.5, and SRES, as shown in Figs. 5, 6, 7, 8 and 9, respectively. The experiment predicts the mean impacts in three future time slices 2020–2039, 2040–2069, and 2070–2100 (Fig. 10).

Table 3 The parameters in the dataset used

Parameter	Description
Crop	Three strategic Crops (Maize, Rice, Wheat)
Current Average Temperature (dC) area weighted °C	Current average annual temperature (2001–2010) from the 0.5°-grid data. For gridded crop simulations, growing areas in each grid/crop were weighted to derive area-weighted averages
Current Annual Precipitation (mm) area weighted Mm	Current annual precipitation (2001–2010) from the 0.5°-grid data. For gridded crop simulations, growing areas in each grid/crop were weighted to derive area-weighted averages
Future Mid-point	Mid-year of the projected period
Baseline Mid-point	Mid-year of the baseline period
Time slice	(2020–2039), (2040–2069), (2070–2100)
Climate scenario	Climate scenarios, such as "RCP2.6", "RCP4.5", "RCP6.0", "RCP8.5" and SRES
Annual precipitation change (mm)	Annual precipitation changes from the Baseline-Mid-point year to the Future-Mid-point year
Projected yield(t/ha)	Projected yield from each study
Climate impacts %	Relative yield change by climate change
Climate impacts per °C % °C^{-1}	Relative yield change per degree
Climate impacts per decade % 10 yr^{-1}	Relative yield change per decade
CO_2 ppm	Mean CO2 concentration used for simulation
Fertilizer	Two categories "Yes" or "No"
Irrigation	Two categories "Yes" or "No"
Cultivar	Two categories "Yes" or "No"
Soil organic matter management	Two categories "Yes" or "No"
Planting time	Two categories "Yes" or "No"
Tillage	Two categories "Yes" or "No"
Adaptation	Two categories "Yes" or "No"
Adaptation type	Nine adaptation measures (No, Fertiliser, Irrigation, Cultivar, Soil Organic matter, Planting time, Tillage, Others, Combined)

Table 4 Confusion matrix

	Predicted as positive	Predicted as negative
Positive	True Positives (TP)	False Negatives (FN)
Negative	False Positive (FP)	True Negatives (TN)

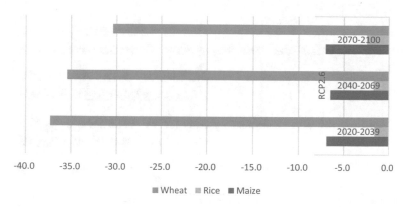

Fig. 5 Means of the projected impacts on three strategic crops on RCP2.6 climate scenario

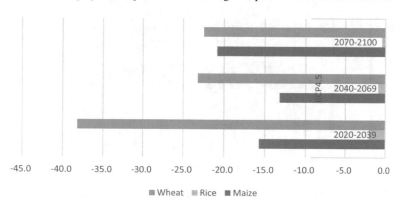

Fig. 6 Means of the projected impacts on three strategic crops on RCP4.5 climate scenario

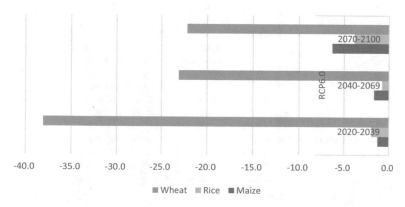

Fig. 7 Means of the projected impacts on three strategic crops on RCP6.0 climate scenario

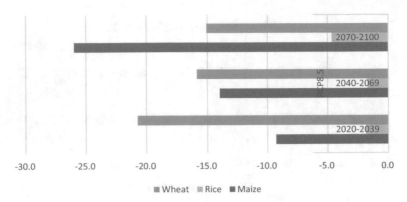

Fig. 8 Means of the projected impacts on three strategic crops on RCP8.5 climate scenario

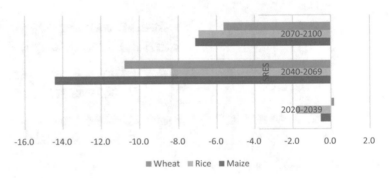

Fig. 9 Means of the projected impacts on three strategic crops on SRES climate scenario

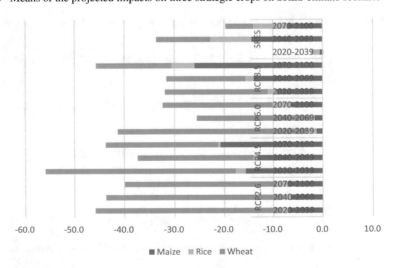

Fig. 10 Means of the projected impacts on three strategic crops in different climate scenarios

The results obtained from this experiment show that the wheat crop is the most affected crop by the climate change data compared to maize and rice, in the different climate scenarios in the near and far future.

5.4 Experiment 2: The Accuracy Values for the Proposed ACRM Compared to Different Models

The performance of the proposed ACRM is compared to the performance of MLR [33], J48 [34], Random Tree, REPTree [34], and ANN [35] models in terms of the accuracy of recommending the suitable crop, as shown in Table 5 and Fig. 11.

From the accuracy results, it can be concluded that ACRM performs better than other models in the three targeted crops.

Table 5 The accuracy values for the proposed ACRM compared to different models

Model	Wheat (%)	Maize (%)	Rice (%)
The proposed ACRM	**98.2**	**98.7**	**98.1**
ANN	96.8	97.5	95.71
MR	95.4	89.8	88
J48	96.1	91.8	89.9
Random tree	91.2	85.9	83.8
REPTree	89.7	83.9	82.5

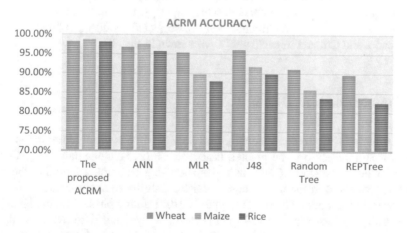

Fig. 11 The Accuracy values for the proposed ACRM

The Most Significant Parameters

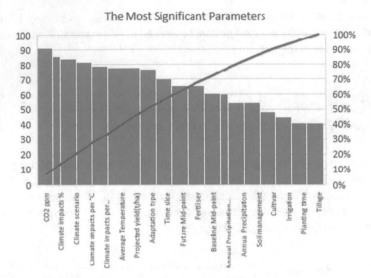

Fig. 12 The most significant climate change factors affected the crop recommendation decision

5.5 Experiment 3: The Most Significant Climate Change Factors that Affected Crop Recommendation

This experiment is used to identify the most significant Climate Change factors that affected the Crop Recommendation decision. The proposed ACRM detected the top most frequent features in the dataset, which are CO_2, Climate impacts, Adaptation type, Average Temperature, Climate scenario, Projected yield(t/ha), Climate impacts per decade, and Climate impacts per °C, as shown in Fig. 12.

6 Conclusion and Discussion

An Automated Crop Recommendation Model (ACRM) is proposed to recommend the most suitable crop to be planted based on Climate change data using Convolutional Neural Network (CNN) as a deep learning technique. Utilizing the deep learning technique in the model bounces the capability to analyze a large number of features and examine diverse data types for the training phase. The model targets three strategic crops growing in Egypt (Maize, Wheat, and rice), which will help the decision makers choose a suitable choice based on analyzed data. An optimized Conventional Neural Network is implemented in the crop recommendation model for predicting the most influential factors that affect crop yield production using the Grey Wolf Optimization algorithm (GWO).

Unlike other pre-trained models, this model is trained from scratch using different layers. The proposed model helps the decision makers to choose the right crop during the cropping season based on the climate change data and its impacts on the crops. From the results obtained, it is discovered that the Wheat crop is the crop that will be affected by the climate change data in the near and far future.

The results from ACRM experiments show the superiority of ACRM over other recommendation models using the three targeted crops, as the accuracy obtained from ACRM was 98.2, 98.7, and 98.1% for Wheat, Maize, and rice crops, respectively. In addition, the ACRM model The proposed ACRM detected the top most frequent features in the climate change data that affect crop yield production: the CO_2, Climate impacts, Adaptation type, Average Temperature, and the Climate scenario.

References

1. FAO. (2000). Irrigation and Drainage Paper I Crop Evapotranspiration (p. 326 https://doi.org/10.1016/S0141-1187(05)80058-6.
2. Tilman, D., Balzer, C., Hill, J., & Befort, B.L. (2011). Global food demand and the sustainable in intensification of agriculture. Proceedings of National Acad. Sci. 108 (50), 20260–20264; Wolfert, S., Ge, L., Verdouw, C., & Bogaardt, M.J. (2017). Big data in smart farming—a review. *Agricultural Systems., 153*, 69–80. Available from: https://doi.org/10.1016/j.agsy.2017.01.023.
3. WMO 2019 Global Climate in 2015–2019: Climate Change Accelerates. Available from: https://public.wmo.int/en/media/press-release/global-climate-2015-2019-climate-change-accelerates.
4. State of the Global Climate Observing System 2021 I World Meteorological Organization (wmo.int) Available from: https://public.wmo.int/en/media/news/state-of-global-climate-obs erving-system-2021.
5. El-Ghamrawy, S.M., El-Desouky, A.I., & Sherief, M. (2009, Mar). Dynamic ontology mapping for communication in distributed multi-agent intelligent system. In *2009 International Conference on Networking and Media Convergence* (pp. 103–108). IEEE.
6. Jha, K., Doshi, A., Patel, P., & Shah, M. (2019). A comprehensive review on automation in agriculture using artificial intelligence. *Artificial Intelligence in Agriculture, 2*, 1–12. Available from: https://doi.org/10.1016/j.aiia.2019.05.004.
7. Elghamrawy, S.M., Hassnien, A.E., & Snasel, V. (2021). Optimized deep learning-inspired model for the diagnosis and prediction of COVID-19. Cmc-Computers Materials and Continua (pp. 2353–2371).
8. Rajak, R.K., Pawar, A., Pendke, M., Shinde, P., Rathod, S., & Devare, A. (2017). Crop recommendation system to maximize crop yield using machine learning technique. *International Research Journal of Engineering and Technology, 4*(12), 950–953.
9. El-Ghamrawy, S.M., & Eldesouky, A.I. (2012). An agent decision support module based on granular rough model. *International Journal of Information Technology and Decision Making, 11*(04), 793–820.
10. Abdelmageed, K., Xu-Hong, C., De-Mei, W., Yan-Jie, W., & Yu-Shuang, Y. (2019). Evolution of varieties and development of production technology in Egypt wheat: A review. *Journal of Integrative Agriculture, 18*, 483–495. https://doi.org/10.1016/S2095-3119(18)62053-2
11. Gaitán, C.F. (2020 Jan 1). Machine learning applications for agricultural impacts under extreme events. In Climate extremes and their implications for impact and risk assessment 2020 Jan 1 (pp. 119–138). Elsevier.
12. Crane-Droesch, A. (2018). Machine learning methods for crop yield prediction and climate change impact assessment in agriculture. *Environmental Research Letters, 13*(11). Available from: https://doi.org/10.1088/1748-9326/aae159.

13. Mehnatkesh, A, Ayoubi, S., Jalalian, A., & Dehghani, A. (2012). Prediction of rainfed wheat grain yield and biomass using artificial neural networks and multiple linear regressions and determination the most factors by sensitivity analysis. In: Information Technology, Automation and Precision Farming. International Conference of Agricultural Engineering—CIGR-AgEng 2012: Agriculture and Engineering for a Healthier Life (p. 1554).

14. Kumar, A., Sarkar, S., & Pradhan, C. (2019) Recommendation system for crop identification and pest control technique in agriculture. In: 2019 International Conference of Communication Signal Processing (pp. 185–189).

15. Ransom, C.J., Kitchen, N.R., Camberato, J.J., Carter, P.R., Ferguson, R.B., & Fernández, F.G., et al. (2019). Statistical and machine learning methods evaluated for incorporating soil and weather into corn nitrogen recommendations. *Computers and Electronics in Agriculture 164*(104872). Available from: https://doi.org/10.1016/j.compag.2019.104872.

16. Duraisamy, V., Singh, S.K., Nisha, S., Pramod, T., Chandran, P., & Duraisami, V.P., et al. (2017). Assessment of spatial variability of soil properties using geospatial techniques for farm level nutrient management. *Soil Tillage Research, 169*, 25–34. https://doi.org/10.1016/j.still.2017.01.006

17. Noya, I., González-García, S., Bacenetti, J., Fiala, M., & Moreira, M.T. (2018). Environmental impacts of the cultivation-phase associated with agricultural crops for feed production. *Journal of Cleaner Production, 172*, 3721–3733.

18. Tebaldi, C., & Lobell, D. (2018). Estimated impacts of emission reductions on wheat and maize crops. *Climate Change, 146*, 533–545.

19. Bonan, G.B., & Doney, S.C. (2018). Climate, ecosystems, and planetary futures: The challenge to predict life in Earth system models. *Science 359*, eaam8328.

20. Kurepin, L.V., Ivanov, A.G., Zaman, M., Pharis, R.P., Hurry, V., & Hüner, N.P. (2017). Interaction of glycine betaine and plant hormones: Protection of the photosynthetic apparatus during abiotic stress. In Photosynthesis: Structures, Mechanisms, and Applications (pp. 185–202). Springer.

21. Dong, H., Bai, L., Chang, J., & Song, C.-P. (2018). Chloroplast protein PLGG1 is involved in abscisic acid-regulated lateral root development and stomatal movement in Arabidopsis. *Biochemical and Biophysical Research Communications, 495*, 280–285.

22. Kuromori, T., Seo, M., & Shinozaki, K. (2018). ABA transport and plant water stress responses. *Trends in Plant Science, 23*, 513–522.

23. Henderson, B., Cacho, O., Thornton, P., van Wijk, M., & Herrero, M. (2018). The economic potential of residue management and fertilizer use to address climate change impacts on mixed smallholder farmers in Burkina. *Faso. Agricultural System, 167*, 195–205.

24. Barnabás, B., Jäger, K., & Fehér, A. (2008). The effect of drought and heat stress on reproductive processes in cereals. *Plant, Cell and Environment, 31*, 11–38.

25. Wang, Z., & Huang, B. (2004). Physiological recovery of Kentucky bluegrass from simultaneous drought and heat stress. *Crop Science, 44*, 1729–1736.

26. Xu, Z.Z., & Zhou, G.S. (2006). Combined effects of water stress and high temperature on photosynthesis, nitrogen metabolism and lipid peroxidation of a perennial grass Leymus chinensis. *Planta, 224*, 1080–1090.

27. Nakicenovic, N., Alcamo, J., Davis, G., Vries, B.D., Fenhann, J., Gaffin, S., Gregory, K., Grubler, A., Jung, T.Y., Kram, T., & La Rovere, E.L. (2000). Special report on emissions scenarios.

28. "Representative Concentration Pathways (RCPs)". IPCC. Retrieved 13 February 2019.

29. Van Vuuren, D.P., Edmonds, J., Kainuma, M., Riahi, K., Thomson, A., Hibbard, K., Hurtt, G.C., Kram, T., Krey, V., Lamarque, J. F., & Masui, T. (2011). The representative concentration pathways: an overview. *Climatic change, 109*(1), 5–31.

30. Moss, R., et al. (2008). *Towards new scenarios for analysis of emissions, climate change, impacts, and response strategies (PDF)* (p. 132). Intergovernmental Panel on Climate Change.

31. Weyant, J., Azar, C., Kainuma, M., Kejun, J., Nakicenovic, N., Shukla, P.R., La Rovere, E., & Yohe, G. (2009 Apr). Report of 2.6 Versus 2.9 W/m^2 RCPP Evaluation Panel (PDF). IPCC Secretariat.

32. Hasegawa, T., Wakatsuki, H., Ju, H., Vyas, S., Nelson, G.C., Farrell, A., Deryng, D., Meza, F., & Makowski, D. (2022). A global dataset for the projected impacts of climate change on four major crops. *Scientific data, 9*(1), 1–11.
33. Cho, S., & Lee, Y. (2019). "Deep learning-based analysis of the relationships between climate change and crop yield in China." *XLII*(September), 93–95.
34. Munandar T.A. (2017). The classification of cropping patterns based on regional climate classification using decision tree approach. https://doi.org/10.3844/jcssp.2017.408.415
35. Madhuri, J., & Indiramma, M. (2021). Artificial neural networks based integrated crop recommendation system using soil and climatic parameters. *Indian Journal of Science and Technology, 14*(19), 1587–1597.

Cost Effective Decision Support System for Smart Water Management System

Amany Magdy Mohamed, Ashraf Darwish, and Aboul Ella Hassanien

1 Introduction

Cities are heavily vulnerable to climate change; they are affected by frequent extremes in heat and cold, storms and cyclones, and rising sea levels. However, they make up a significant portion of the global greenhouse gas (GHG) emissions—roughly 72%—and are a key cause of climate change. The cities in the developing world will experience great change because of urban population growth. They need to deal with that change effectively, considering the limitations of the resources and institutional capabilities, which create tough challenges. Due to these difficulties, urban areas have evolved into complex social ecologies where ensuring sustainability and a high standard of living are crucial [1]. Cities are consequently compelled to look for the best solutions for many important issues, including sustainable development, public services, energy, education, the environment, and safety.

Cities can be described as a system of systems. They involve many systems, including transport, health and biodiversity, food, energy, water and sewerage, and cultural, social, and economic systems. Consequently, solving city problems is very complex as they are interconnected, and solving one problem in a specific system can have unintended consequences in another system. Thus, instead of solving these problems individually, they should be solved using an integrated approach [2–4].

To handle the complexity of urban living and implement solutions for diverse city problems, the Smart City idea seeks to address the issues caused by the rapid rate of population expansion in urban areas. Yet, there is no standardized definition of "smart

A. M. Mohamed (✉) · A. E. Hassanien
Faculty of Computers and Artificial Intelligence, Cairo University, Cairo, Egypt
e-mail: amanymagdy@fci-cu.edu.eg

A. Darwish
Faculty of Science, Helwan University, Cairo, Egypt

A. M. Mohamed · A. Darwish · A. E. Hassanien
Scientific Research Group in Egypt (SRGE), Cairo, Egypt

© The Author(s), under exclusive license to Springer Nature Switzerland AG 2023　　　　207
A. E. Hassanien and A. Darwish (eds.), *The Power of Data: Driving Climate Change with Data Science and Artificial Intelligence Innovations*, Studies in Big Data 118,
https://doi.org/10.1007/978-3-031-22456-0_12

cities." However, this term is used to solve cities' sustainability challenges based on data and smart technologies and combine them with innovation and creativity to develop innovative solutions to citizens' needs and urban challenges. This term is related to the present and the future; many cities are in the process of making themselves smart, while others are being built to be smart from the start [5, 6].

Climate change significantly impacts water resources; it affects the quality and availability of global freshwater supplies and increases the likelihood of droughts and flooding due to temperature and weather fluctuations [7]. Water resources can be any resources that are useful or potentially useful to humans. 3% of the water on the Earth is fresh water, and 97% is salt water. The demand for high-quality water will increase because of the world's population growth and climate change. Consequently, new water management strategies must be applied to optimize water consumption and reduce the environmental effects of water use on the environment [8].

Making decisions about how to use water resources is the process of water management. Earth's water supply is finite; hence it must be managed sustainably to be protected. Water security should be ensured by future water resource planning and management based on resilience, adaptation, and capacity building. Planning, developing, allocating, and administering the best possible use of water resources—both in terms of quantity and quality—to maximize the resulting economic and social welfare in a fair way and without compromising the sustainability of critical ecosystems. The building of such an integrated water management system is made simple by the idea of smart cities [9].

Smart governance, smart living, smart environments, smart mobility, smart economies, and smart people are the six essential components of a smart city system. Smart water is a crucial part of the smart environment. The idea of smart cities inspires the idea of it. A smart city's endowment of smart infrastructure makes it easier for smart water to exist because it is used for many of the resident's needs, including water [9].

Before the development of smart cities, smart water management was not widely used. This chapter proposes an integrated, flexible, and cost-effective DSS for solving the smart water management system's decision problems. It is a highly responsive, intelligent digital system that combines with humans to identify water-related problems and even automatically uses artificial intelligence to solve them in real-time without human intervention. Based on this definition, two types of DSS can be used to solve water-related issues, the model-driven DSS for the decision problems modeled by humans and the data-driven DSS for the decision problems that can be modeled automatically without human interventions.

The rest of this chapter is organized as follows: Sect. 2 provides urbanization trends, while Sect. 3 deals with the main concept of smart cities. Section 4 discusses the proposed DSS. Finally, the conclusion is given in Sect. 5.

2 Urbanization Trends

Urbanization is a complicated socio-economic process that modifies the built environment, turning once rural settlements into urban ones and redistributing a population's geographic distribution from rural to urban areas. It affects the prevailing occupations, way of life, culture, and behavior, changing urban and rural areas' social and demographic structures. Increases in the number, size, and population of urban settlements and in the proportion of urban to rural citizens are one of the main effects of urbanization. The world's population is predicted to rise from 7.63 billion in 2018 to 9.77 billion in 2050, with urban areas seeing the majority of this growth. The urban population is projected to rise from 4.2 billion to 6.7 billion [10].

The six geographical categories of Africa, Asia, Europe, Latin America and the Caribbean, Northern America, and Oceania are used by the United Nations [10] to classify countries and regions. The vastly differing experiences among the geographic regions that make up those categories are the cause of the distinct patterns of urbanization that have been seen in the more developed regions and the less developed regions. Less developed regions will have a substantially higher percentage of the population living in urban areas than highly developed ones. There are notable differences in the urbanization trends between less developed and more developed regions. Currently, fewer than half of the population in less developed regions lives in rural areas, whereas the vast majority of people in more developed regions live in urban areas. However, the less developed regions' proportion of the global urban population has been increasing due to their urban population developing far more quickly than that of the more developed regions, as in Fig. 1.

According to Fig. 2, the urban population in Asia, Europe, Latin America, the Caribbean, and North America will decline in 2050 compared to 2018, whereas it will rise in Africa. However, as depicted in Fig. 3, Asia will have the highest urbanization rates. More than half of the expected rise in the world's population up to 2050 will be centered in Africa, particularly Egypt, one of the nine African countries with a level of urbanization of more than 25%. Alexandria is Egypt's second-largest city by urban population after Cairo, which corresponds to the Governorate of Al-Qahirah and has a population of more than 10 million [10].

3 Smart Cities Concept

Increased demand for natural resources like water and energy, increased pollution, and effects on biodiversity are all results of rapid urbanization and urban population growth. Cities presently account for between 67 and 76% of global energy consumption and between 71 and 76% of CO_2 emissions [10]. There is a large population density, poor sanitary conditions, and high water contamination levels. Cities produce 1.7–1.9 billion tons of waste, or 46% of the world's waste. Smart urban planning and

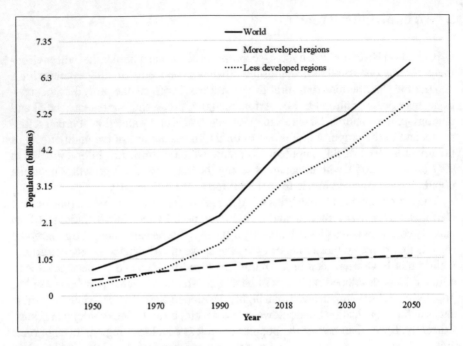

Fig. 1 Distribution of urban population

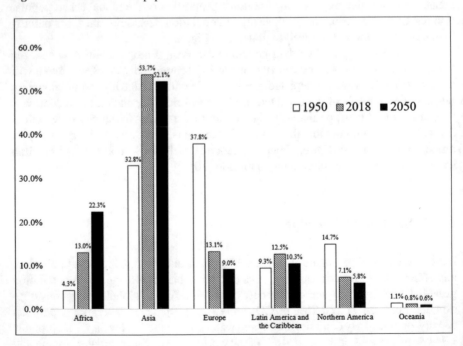

Fig. 2 Urban population in the six geographic regions

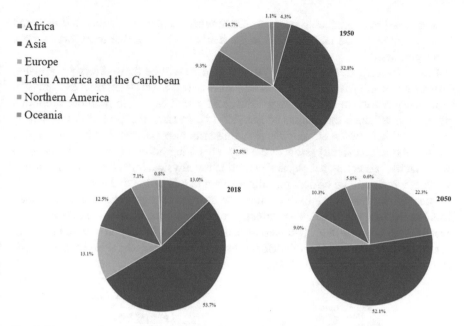

Fig. 3 The proportion of urban residents in each of the six geographic areas

understanding urban population trends can help minimize the adverse urbanization impacts and environmental degradation [10].

Consequently, the concept of smart cities becomes necessary and a natural strategy to mitigate these challenges. Despite of the cost of building a smart city, it can reduce carbon emissions, energy consumption, transportation requirements, water consumption, and city waste.

Cities are encouraged to become "smart cities" to solve the challenges of rapid urbanization, increased pressure on city services, and climate change. Smart cities do not have a standard definition and it is defined with different perspective and views ranging from purely technological to focusing on achieving sustainability [11].

Enhancing city functions requires more than ICT access for large-scale data collection and analytics, and focusing only on these criteria can paint an incomplete picture of smart cities. Some definitions focus on technology only and neglect sustainability or social capital, which increases the impact of the climate change mitigation and adaptation efforts [12, 13]. Some definitions emphasize collecting large-scale data and harnessing big data analytics to improve urban performance and city functions [14, 15].

Consequently, the social and economic aspects need to be described in the smart cities context. In [16], six key characteristics of smart cities are outlined: smart people, smart living, smart governance, smart economy, smart environment, and smart mobility. These characteristics encompass social/human capital, community participation, quality of life, and the city's competitiveness within a global or regional market. Sustainable resource use, environmental protection, and the need for ICT infrastructure and sustainable, available, and accessible transportation are also highlighted.

Efforts to mitigate and adapt to climate change will be more directly correlated with the developing trend of smart, sustainable cities. Several definitions emphasize community involvement, ICT integration, and social capital to promote economic growth and enhance the quality of life [17–19]. In [20], the definition focused on the integration between the tangible (i.e., infrastructure systems) and intangible (i.e., social and human capital) assets of the city. This integration involves building adaptive capacity, protecting people, and preparing the physical infrastructure for extreme weather, which will be critical for climate change mitigation and adaptation.

None of those above definitions mention climate change mitigation and adaptation, but in [21], smart cities are directly connected with sustainable development. Based on the six identified characteristics, the smart city is built on a smart combination. Each character has a set of attributes that can be formed to tackle city challenges. The characteristics and their attributes are shown in Fig. 4.

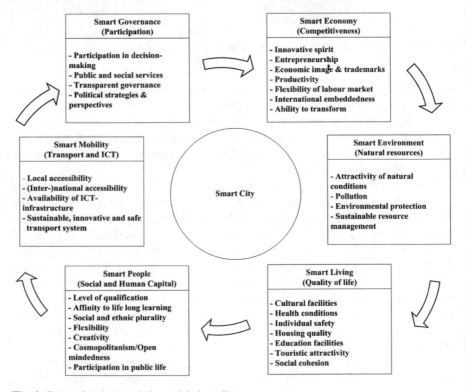

Fig. 4 Smart city characteristics and their attributes

4   The Proposed DSS for the Smart Water Management System

Smart water management was not widely employed before the creation of smart cities. The smart water management system is an integrated system that merges the decision-making problems related to managing different categories such as water resources, natural drainage, wastewater, water supply, flood and inundation, and aging water infrastructure. It is described as a highly responsive, intelligent digital system that collaborates with people to identify water-related issues. It even automatically employs artificial intelligence to remedy those issues in real-time without human involvement. Based on this definition, two types of DSS can be used to solve water-related issues, the model-driven DSS for the decision problems modeled by humans and the data-driven DSS for the decision problems that can be modeled automatically without human interventions.

This chapter proposes an integrated, flexible, and cost-effective DSS for solving the smart water management system's decision problems. The proposed DSS is cloud-based DSS that consists of a group of DSS. Each DSS contains the decision problems related to one smart water management system category. For example, based on the mentioned categories, the proposed DSS consists of six different DSS.

4.1   Decision Support System

With the growing uncertainty and complexity in many decision situations, the decision maker (DM) forced a great challenge to make the right decision at the right time and with minimum cost. DSS is a computer-based system built to help DM to make better and more effective decisions, especially for ill-structured or weakly-structured problems. The capabilities of DSS can be increased by using technological innovation.

Power D. J. defines a DSS as an interactive computer-based system or subsystem designed to assist decision-makers in using communications technologies, data, documents, knowledge, and/or models to identify and resolve issues, carry out tasks related to the decision-making process, and make decisions. Any computer system that improves the capacity of an individual or group to make decisions is referred to as a "DSS." Generally speaking, DSS is a subset of computerized information systems that assist with decision-making activities [22].

Cloud computing is a model for delivering computing resources as services with automation, flexibility in payment, a high degree of elasticity, and, as a result, lower cost [23]. The National Institute of Standards and Technology (NIST) defines cloud computing formally as a model for enabling universal, convenient, on-demand network access to a shared pool of reconfigurable computing resources (e.g., networks, servers, storage, applications, and services) that can be quickly provisioned and released with little management work or service provider interaction. This cloud model consists of three service models (Software as a Service (SaaS), Platform as a Service (PaaS), and Infrastructure as a Service (IaaS), and five key characteristics (on-demand self-service, broad network access, resource pooling, rapid elasticity, and measured service), and four deployment models (private cloud, community cloud, public cloud, hybrid cloud) [24, 25].

To gain the full utilization of cloud computing capabilities and advantages, the system/application needs to build as a cloud-native system/application; its functionality is implemented as the composition of services running on the cloud.

DSS has three main components: (1) user interface, (2) models and analytical tools, and (3) database. The model (simulation, statistical, logic, and optimization, ...) is the main component that differentiates DSS from other management information systems (MIS). The functionality of DSS is provided in the model component. The decision problem is represented by a suitable model and the solutions of this model are called decision alternatives. Based on the type of the DSS, the model can be defined a priori of the decision-making process or during it [26, 27].

In this chapter, two types of DSS are discussed; the data-driven DSS and the model-driven DSS. In data-driven DSS, the model need not be constructed a priori, rather it is defined and constructed through the analysis of a large amount of data. While, in model-driven DSS, the model is assumed to be existed before decision making activities. It is not usually data intensive. To obtain decision alternatives, data and user input are used to calculate model parameters.

4.2   Cloud-Native Systems

To determine the usage of the term "cloud-native" in research, the term was explored from Google trends from 2006—the birth of the cloud—to 2021. According to Fig. 5, in the beginning, the term was used quite frequently, then over the following years, its usage of it decreased. However, since 2015, the term is more and more frequently used again [29]. Kratzke and Qunit—after 10 years of cloud computing – provided a systematic mapping study to analyze research papers covering "cloud-native" topics, research questions, and engineering methodologies. Based on this study, they defined cloud-native applications (CNA) as a distributed, elastic, and horizontally scalable system built of (micro)services that isolate a state in a minimum of stateful components. The application and every standalone deployment unit were created using cloud-focused design patterns and run on an elastic, self-service platform. [29].

With microservice architecture, each service has its database. The services are communicated via an application programming interface (API) to ensure service separation and avert the risk of tight coupling. As a result, each service can scale, change, and deploy independently without requiring too much sharing with other services. Besides, it can be isolated quickly, and the rest of the system continues working.

The microservice architecture for an application is created through two steps: (1) define the service and (2) decide how the service collaborates. In this chapter, we

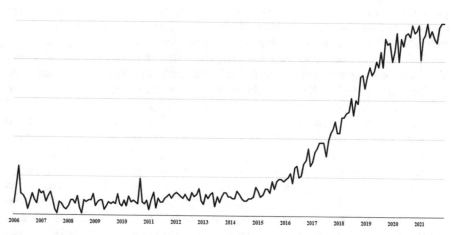

Fig. 5 Google trends (01.01.2006 until 01.11.2021) of the term "cloud-native"

focus on defining the services. There are different strategies to define services, but in this chapter, we concentrate on two of them; the business capability strategy and the single responsibility principle (SRP) strategy [30, 31].

In business capability strategy, the services are identified based on the business capability concept. This concept is embraced from business architecture modeling to capture what business can do. Organizing and identifying services around business capabilities make the architecture stable because they are stable despite changing the way of doing them. At the same time, the SRP strategy is based on the cohesion concept. Robert C. Martin defines this concept as "Gather together those things that change for the same reason, and separate those things that change for different reasons." [32]. The services identified according to this strategy are cohesive and small services with a single responsibility and hence a single reason to change.

4.3 The Proposed DSS for Smart Water Management System

A smart water system consists of water supply, water resources, wastewater, natural drainage, and aging water infrastructure. Each component has its decision-making problems. Although each component merges the decision problems related to one purpose, each problem has its characteristics that lead to a different model to represent it and, as a result, a different solution approach to obtain the decision alternatives. The smart water system should be an integrated system to fully optimize the whole system because solving the component's problems separately may negatively affect the other components.

For example, the wastewater components contain different decision models that generally aim to optimize wastewater usage to increase clean water and decrease the impact of wastewater on the environment. This aim is achieved through different activities such as treating, disposing, and reusing wastewater. However, wastewater has different types, like domestic wastewater and industrial wastewater. Each type has characteristics and features that lead to a different model to identify its contaminants.

The proposed DSS is flexible and cost-effective as the concept of a cloud-native system drives it. With microservice architecture, the proposed DSS is developed as a suite of independently deployable services modeled around the interconnected decision problems and communicate with lightweight mechanisms. The services are defined based on the mentioned strategies in the "Cloud-native systems" section.

Each DSS represents a service, in addition to the main service that is responsible for managing, coordinating, and connecting among the services. Each service has its data that make it an independently deployable service. As shown in Fig. 6, the main service is called a "smart water management system" that is responsible to manage, coordinating and connect the other service such as "smart wastewater DSS", "smart water resources DSS" and "smart aging water infrastructure DSS".

The proposed DSS is flexible in dealing with different decision problems and solutions. Besides, the flexibility of using different technologies inside each service and adopting new technology more quickly. The proposed DSS is resilient; with

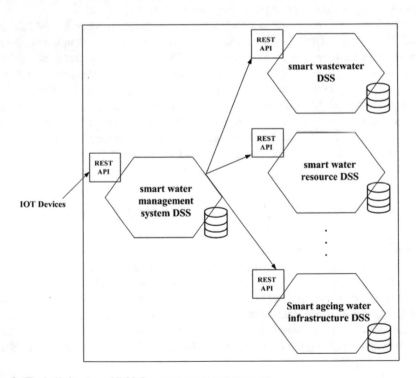

Fig. 6 The holistic view of DSS for smart water management

microservice architecture, the problem can be isolated if one component fails, and other components will continue to work normally. The proposed DSS is independently scalable; each service can be deployed on hardware that's best suited to its resource requirements. Besides ease of deployment, if a change is done to a single service, we can deploy it independently of the rest of the system.

Each service in Fig. 6 except the main service, can be split into a group of services. Each service is defined as a model of the specific decision problem. As shown in Fig. 7, the "smart wastewater management DSS" is expanded to contain all the potential related decision problems. For example, the "treatment and disposal of domestic wastewater" and "recycling and reuse of domestic wastewater". Each problem is defined as an autonomous service because of the difference like each problem which leads to the different decision model. In Fig. 8. The "smart water resources management DSS" is expanded to its related decision problems. For example, the problem of finding the optimal design of flexible water distribution and the problem of forecasting water demand. In Fig. 9, the "smart aging water infrastructure" is expanded to its related decision problems. For example, the problem of monitoring aging infrastructure and the problem of detecting and locating leakages in the water supply network.

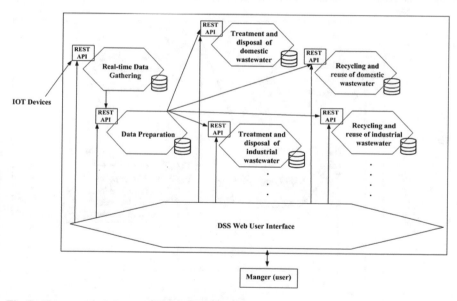

Fig. 7 The expanded view of DSS for smart wastewater

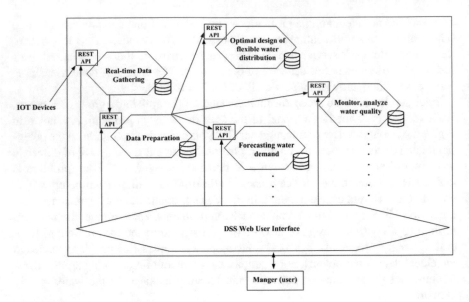

Fig. 8 The expanded view of DSS for smart water resources

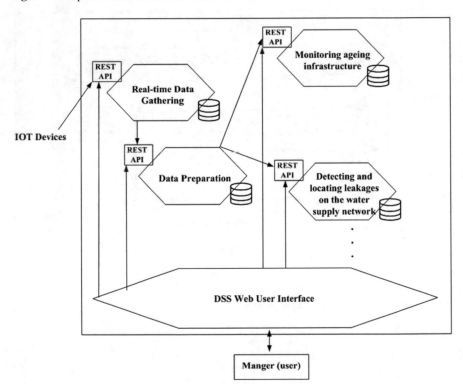

Fig. 9 The expanded view of DSS for smart aging water infrastructure

5 Conclusion

The smart water management system is one of the essential components of the smart environment that optimize water consumption and reduce the environmental effects of water use on the environment. This chapter proposed an integrated, flexible, and cost-effective DSS for solving the decision problems of the smart water management system. The proposed DSS is cost-effective as the concept of a cloud-native system drives it. Besides, it is flexible in dealing with different decision problems and solutions.

References

1. Arroub, B. Zahi, E. Sabir, & Sadik, M. (2016). A literature review on Smart Cities: Paradigms, opportunities and open problems. In *Proceedings of International Conference on wireless networks and mobile communications (WINCOM)*, Fez (pp. 180–186). https://doi.org/10.1109/WINCOM.2016.7777211
2. Gade, D. (2019). Introduction to smart cities and selected literature review. *International Journal of Advance and Innovative Research, 6*(2), 7–15.
3. Mohanty, S.P., Choppali, U., & Kougianos, E. (2016). Everything you wanted to know about smart cities: The internet of things is the backbone. *IEEE Consumer Electronics Magazine, 5*(3), 60–70. https://doi.org/10.1109/MCE.2016.2556879
4. Kozłowski, W., & Suwar, K. (2021). Smart City: Definitions, dimensions, and initiatives. *European Research Studies, 24*(3), 509–520.
5. Patrão, C., Moura, P., & Almeida, A.T.D. (2020). Review of smart city assessment tools. *Smart Cities, 3*(4), 1117–1132. https://doi.org/10.3390/smartcities3040055
6. Adiyarta, K., Napitupulu, D., Syafrullah, M., Mahdiana, D., & Rusdah, R. (2020). Analysis of smart city indicators based on prisma: Systematic review. *IOP Conference Series: Materials Science and Engineering, 725*(1), 012113.
7. Kim, K.G. (2018). *Low-carbon smart cities*. Springer.
8. Gonçalves, R., Soares, J.M., & Lima, R.M.F. (2020). An IoT-based framework for smart water supply systems management. *Future Internet 12*(7). https://doi.org/10.3390/fi12070114.
9. Kumar, V., Mohammed Firoz, C., Bimal, P., Harikumar, P.S., & Sankaran, P. (2020). Smart water management for smart Kozhikode metropolitan area. *Smart Environment for Smart Cities* (pp. 241–306). Springer.
10. UN Department of Economic and Social Affairs Population Division, World Urbanization Prospects: The 2018 Revision.
11. Obringer, R., & Nateghi, R. (2021). What makes a city 'smart' in the Anthropocene? A critical review of smart cities under climate change. *Sustainable Cities and Society, 75*, 103278. https://doi.org/10.1016/j.scs.2021.103278
12. Batty, M., Axhausen, K.W., Giannotti, F., Pozdnoukhov, A., Bazzani, A., Wachowicz, M., Ouzounis, G., & Portugali, Y. (2012). Smart cities of the future. *The European Physical Journal Special Topics, 214*(1), 481–518. https://doi.org/10.1140/epjst/e2012-01703-3
13. Nam, T., & Pardo, T.A. (2011). Conceptualizing smart city with dimensions of technology, people, and institutions. In: *Proceedings of the 12th annual international digital government research conference: Digital government innovation in challenging times* (pp. 282–291). Association for Computing Machinery. https://doi.org/10.1145/2037556.2037602.
14. Kitchin, R. (2014). The real-time city? Big data and smart urbanism. *GeoJournal, 79*(1), 1–14. https://doi.org/10.1007/s10708-013-9516-8

15. Marsal-Llacuna, M.L., Colomer-Llin'as, J., & Mel'endez-Frigola, J. Lessons in urban monitoring taken from sustainable and livable cities to better address the smart cities initiative. *Technological Forecasting and Social Change 90*, 611–622. https://doi.org/10.1016/j.techfore.2014.01.012.

16. Giffinger, R., Fertner, C., Kramar, H., & Meijers, E. (2007). City-ranking of European medium-sized cities. *Cent. Reg. Sci. Vienna UT, 9*(1), 1–12.

17. Deakin, M., & l Waer, H.A. (2011). From intelligent to smart cities. *Intelligent Buildings International 3*(3), 140–152. https://doi.org/10.1080/17508975.2011.586671.

18. Caragliu, A., Bo, C.D., & Nijkamp, P. (2011). Smart cities in Europe. *Journal of Urban Technology, 18*(2), 65–82. https://doi.org/10.1080/10630732.2011.601117

19. Angelidou, M. (2015). Smart cities: A conjuncture of four forces. *Cities, 47*, 95–106. https://doi.org/10.1016/j.cities.2015.05.004

20. Neirotti, P., De Marco, A., Cagliano, A.C., Mangano, G., & Scorrano, F. (2014). Current trends in smart city initiatives: Some stylised facts. *Cities, 38*, 25–36. https://doi.org/10.1016/j.cities.2013.12.010

21. Ahvenniemi, H., Huovila, A., Pinto-Sepp¨a, I., & Airaksinen, M. (2017). "What are the differences between sustainable and smart cities?", *Cities, 60*, 234–245. https://doi.org/10.1016/j.cities.2016.09.009

22. Power, D.J. (2009). *Decision support basics*, Business Expert Press.

23. Aljabre, A. (2012). Cloud computing for increased business value. *International Journal of Business and Social Science 3*(1).

24. Rittinghouse, J.W., & Ransome, J.F. (2016). *Cloud computing: Implementation, management, and security*, CRC press. https://doi.org/10.1201/9781439806814.

25. Mell, P., & Grance, T. (2011). *The NIST definition of cloud computing*, Computer Security Division, Information Technology Laboratory, National Institute of Standards and Technology Gaithersburg, MD 20899–8930

26. Emery, J.C. (1987). *Management information systems: The critical strategic resource*. Oxford University Press Inc.

27. Bell, P.C. (1992). DSSs: Past, present and prospects. *Journal of Decision Systems, 1*(2–3), 127–137. https://doi.org/10.1080/12460125.1992.10511521

28. Kersten, G.E., Mikolajuk, Z., Yeh, A.G.O. (2000). *DSSs for sustainable development: a resource book of methods and applications*. Springer Science and Business Media.

29. Kratzke, N., & Quint, P.C. (2017). Understanding cloud-native applications after 10 years of cloud computing-a systematic mapping study. *Journal of Systems and Software, 126*, 1–16. https://doi.org/10.1016/j.jss.2017.01.001

30. Mark, R. (2015). *Software Architecture Patterns-Understanding Common Architecture Patterns and When to Use Them*, O'Reilly, Febrero.

31. Martin, R.C. (1995). *Designing object-oriented C++ applications*. Prentice Hall.

32. Martin, R.C. (2002). *Agile software development: Principles, patterns, and practices*. Prentice Hall.

The Role of Artificial Intelligence in Water Management in Agriculture for Climate Change Impacts

Wessam El-ssawy, Ashraf Darwish, and Aboul Ella Hassanien

1 Impact of Artificial Intelligence on Environmental Issues

1.1 Global Warming and Artificial Intelligence

AI is now considered a significant study topic to handle the majority of the present environmental sustainability concerns due to the global environmental issues of the twenty-first century [11]. The flood of information has been substantially magnified by the processing of digital data from humans to machines. There has been a considerable advancement in information and communication technology (ICT) that will benefit farmers and stakeholders alike in management applications of digital farming. Promoting fresh technology solutions into remote areas [44]. AI ideas and programming paradigms are used by multi-objective intelligent systems to provide solutions to harmonic optimization issues [45]. Since the development of multi-target particle swarm optimization two decades ago, this category of mathematical models has gained popularity [8].

AI algorithms can alter data and extract useful information from it, which can help people make wise decisions. By introducing new concepts and techniques like machine learning (ML), natural language processing (NLP), machine vision (MV),

W. El-ssawy (✉)
Agricultural Engineering Research Institute (AEnRI), Agricultural Research Center (ARC), Giza, Egypt
e-mail: Eng_Wess50@yahoo.com

A. Darwish
Faculty of Science, Helwan University, Cairo, Egypt

A. E. Hassanien
Faculty of Computers and Artificial Intelligence, Cairo University, Cairo, Egypt

W. El-ssawy · A. Darwish · A. E. Hassanien
Scientific Research Group in Egypt (SRGE), Cairo, Egypt

© The Author(s), under exclusive license to Springer Nature Switzerland AG 2023
A. E. Hassanien and A. Darwish (eds.), *The Power of Data: Driving Climate Change with Data Science and Artificial Intelligence Innovations*, Studies in Big Data 118,
https://doi.org/10.1007/978-3-031-22456-0_13

artificial neural networks (ANN), etc., problem-solving and automation are made very simple. According to Jha et al. [21], ML and ANN are the most often used methodologies in research on automated agriculture. ML algorithms allow for the use of both labeled (supervised learning) and unlabeled (unsupervised learning) data. Follow connected processes using supervised learning. Modern automated farming systems mainly rely on ANN, which performs well in challenging categorization tasks. Biological neurons served as an inspiration for the layer-based construction of the ANN. They study the non-linear complicated relations using geometry. Deep learning-based computer vision technology. It has made enormous progress and performance demonstrated exemplary in image segmentation, discovery, classification, and retrieval-related tasks, thus the interest in reviving the scientific community in ANNs [7, 20]. These ANNs are commonly used in agricultural automation and are typically built based on Convolutional Neural networks (CNN) [7, 20].

ML is widely used in research to evaluate the quality of water by predicting models based on optical water properties. These models are very helpful in reducing time losses and simplifying irrigation system control so that manual irrigation systems can be replaced by automatic irrigation systems. So, the maintenance will be done when it is most appropriate. One of the most crucial benefits of using a regression model is the ability to use a straightforward laser device to measure one optical or physical parameter and then predict the regression of other parameters, saving time and money spent on expensive laser devices to measure the three optical properties (reflection, transmission, and absorption). The models used include support vector machine (SVR), multiple linear regression (MLR), random forest (RF), and XGBoost (XGB) [9]. Contrarily, certain phenomena like climate change, a lack of water, and overuse of fertilizers necessitate a more effective use of resources in the agricultural sector. Under shifting hydro-climatic conditions, agro-ecosystems must be reoriented to achieve rising socio-economic goals while putting less strain on environmental resources [39].

Numerous risks are posed by the effects of climate change, and changes in the amount and quality of water resources and agriculture yield are among them [13]. Several pieces of data that show how much humans are affecting the Earth's climate make climate change more certain than ever. Along with a rise in sea level, a rapid decline in Arctic Sea ice, and other climate-related changes, the atmosphere, and oceans have warmed. The impacts of climate change on people and the environment are becoming more obvious. Wildfires, heat waves, and unprecedented floods have caused billions of dollars in damage. In reaction to shifting patterns of temperature and precipitation, habitats quickly change [37].

Climate change is a progressive alteration of the climate system brought on by both natural and man-made factors. Climate refers to a long-term shift in the state of the atmosphere of a certain place or region. Climate change is the result of intricate interactions between the atmosphere, hydrosphere, biosphere, cryosphere, and lithosphere, which are all components of the climate system [33].

All climate models indicate that the regions where agriculture is produced will see more extreme weather, including more heavy rains, storms, and droughts. Such severe weather conditions will influence when and where illnesses spread, creating serious dangers and possibly leading to crop failure. Because ecological, social, and economic systems in emerging nations are already under pressure from factors including fast population growth, the expansion of industry, and economic inequality, climate change poses an extra burden [28].

Natural and man-made causes make up the majority of the causes of climate change. Changes in solar activity, volcanic eruption, sea water temperature, ice sheet distribution, westerly waves, and atmospheric waves are a few examples of natural causes. On the other side, the artificial causes include activities like the emission of carbon dioxide from industrial and agricultural production, deforestation, acid rain, and the use of Freon to destroy the ozone layer, as well as global warming caused by an increase in greenhouse gas production. Science, Technology, and Education: PACEST [33]. The term "greenhouse effect" refers to phenomena in which atmospheric elements like carbon dioxide and water vapor shield solar energy that is reaching the Earth from the radiation outside the atmosphere, raising the planet's average temperature. The amount of carbon dioxide in the atmosphere could cause the temperature to increase [14].

1.2 Impacts of Climate Change on Production

The biggest factor influencing agricultural productivity is climate change. Crop growth is anticipated to be directly impacted by potential changes in temperature, precipitation intensity, and CO_2. With perfect adaptation and sufficient irrigation amounts, the overall impact of climate change on the world's food supply is seen as low to moderate [18]. The cumulative effect of carbon dioxide fertilization can increase agricultural output worldwide. Climate changes that affect water supplies will also have an impact on agriculture [12].

Positive effects of global warming include an increase in production due to the fertilizing effect caused by an increase in carbon dioxide levels in the atmosphere, an expansion of the areas available for the production of tropical and/or subtropical crops, an increase in two-crop cultivation due to an extension of the planting season, and a decrease in crop damage. Due to the low temperatures, winterization lowers the cost of heating for the growing of crops in protected facilities [25].

The harmful effects of global warming, such as decreased crop quality and quantity due to a shorter growth period after high-temperature levels; Low sugar content, bad coloring, low fruit storage stability; increasing weeds, pests, and insects in yields; reduced fertility of the earth due to the accelerated decomposition of organic matter; and increased soil erosion due to increased precipitation [25]. In Fig. 1 Positive and negative effects of global warming on the agricultural sector [24].

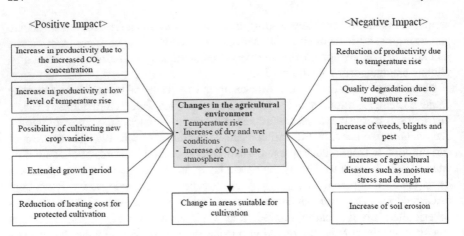

Fig. 1 Potential effects of global warming on the agricultural environment

1.3 Effect of Climate Change on Water Management

Climate change will also have an impact on the water cycle [48]. The possibility of salty interference in rivers and groundwater, which can affect the water's character and its potential for use in the home, industrial, and agricultural applications, will also increase as the sea level rises. Numerous effects of climate change on agriculture [13]. There is now great concern about declining soil fertility, increasing salinity, changing the groundwater table, deteriorating irrigation water quality, and resistance to many pesticides in northwest India [32]. Additional twisting effects may be increased by shifting rates of runoff and groundwater recharge, perturbing water supplies, and changes in capital or technological requirements such as irrigation methods and surface water storage [1]. The typical topographic features of its high mountains cause signs of climate change, such as flash floods, to occur more frequently now and with less time to pass between critical water levels and rainstorms. Precipitation levels are rising, and the upcoming rainy season will bring more frequent rains [41].

Water scarcity is high on the development agenda. In particular, a global map depicting water-scarce regions, with statistics on people in water-scarce basins, and suggestions for addressing water scarcity, provided geographic context and focus on dialogues and development agendas [19].

The current lack of adequate fresh water supplies poses difficulties for the sustainability of field projects, particularly with the use of groundwater for irrigation consumed in large quantities as these nations are characterized by extremely high rates of evaporation and transpiration and soils with low water holding capacity [3].

However, more food production will likely mean that agriculture will require more water from rain and irrigation. Moreover, there are increasing demands from cities, for energy, and on top of that, there are uncertainties caused by climate change. Water is likely to become a major constraint to the food production systems of many of the world's breadbaskets. The main question is whether there will be enough water to grow the required food [19]. Meanwhile, millions of the rural poor can benefit from better water services to support their livelihoods and enhance food security. Sub-Saharan Africa has not mobilized water anywhere near the amount needed to improve agricultural productivity. Water does not play the role it should in poverty reduction and economic growth [19].

Egypt consumes 128% of its available water, with 27% of that amount coming from imported food and other things. According to the Ministry of Water Resources and Irrigation, Egypt would need 20% more water by 2020 to meet demand and population growth [15]. The gap between food production and consumption in the Near East and North Africa region is constantly widening as a result of the annual increases in population growth rates, which also cause a large increase in urbanization and slow growth in local food production. There is a pressing need to boost agricultural production in Egypt using the available resources, which necessitates thorough research and adequate answers to the problems that sustainable agriculture in Egypt faces [15].

The need for water, however, must rise by 20% (15 billion m^3/year) by 2020. Water quantity and quality are mutually exclusive. Since the water quality for each use of water must fall within a certain range. As a result, the current rate of quality degradation will undoubtedly worsen the problem of water shortages or raise the expense (i.e., the need for treatment) of using water at levels anticipated in the 1970s through 2020 [29]. Except for sugar, wheat, and oils, Egypt is self-sufficient in practically all agricultural products. Egypt is now one of the biggest food importers in the world, notwithstanding these outliers. The country's foreign exchange reserves are under a tremendous amount of stress as a result of the country's constantly rising agriculture import bill. Early in the 1970s, imports were more than twice as high, but since the mid-1970s, the balance has become negative and the gap has steadily grown [29].

2 Automation and Traditional Farming for Water Management

Agriculture automation is a big rising and problematic issue for any nation. The world's population has grown at a very quick rate, and as a result, there are more people and a greater demand for food [21]. Farmers must employ toxic pesticides on a large scale to degrade the soil since their traditional methods are insufficient to meet the demands that are growing. The area eventually remains fertile and arid as a result of this's numerous implications on agricultural practices [21]. In protected agriculture, traditional methods are thought to be highly productive, but their relative water use may be high due to runoff and incursion, hence, the efficiency of water use may be low. A good farmer may get the same productivity in the soil as in soilless farming but using soilless save 50–100% of water as a result of losing water from excessive soil irrigation and evaporation from the soil surface. If we consider the production per unit of water applied, yields may be increased significantly by soilless systems over soil-based systems. To decrease water loss during cultivation, a soilless system has been developed from an open system to a closed system [35].

a. Closed Systems of Culture to Save Water

Water scarcity made it necessary to develop ecologically friendly strategies to increase water productivity and crop yields [30]. Reduced water and nutrient losses to the environment and increased water usage efficiency are the immediate benefits of closed systems over open systems. Additionally, because closed systems use a minimal amount of substrate.

b. Open Systems of Culture

Open systems where surpluses of nutrients and water (approximately 25%), like in typical soil farming, are supplied but can become waste (Fig. 1). Many technologies are similar and have been developed by employing a range of inert media such as sand, vermiculite, perlite, and pumice. This technology's attraction can be described as being the same strategy of using soil as a growing medium. The substrate's inertness and substantial water-holding capacity are its two most significant qualities. To avoid plant stress, the substrate must maintain an adequate quantity of nutrients and water [30].

Therefore, ways and techniques which can contribute to improving the efficiency of water use and productivity deserve close study such as the soilless culture technique [3].

3 Artificial Intelligence and Soilless Culture for Facing Climate Change

Soilless culture is a cutting-edge method of growing plants that feed nutrient solutions onto an inert organic or inorganic substrate. It is most likely the most intensive farming approach for commercial greenhouse vegetables, where every resource is utilized to its fullest potential to enhance crop yield. For the production of high-value vegetable crops, several researchers have explored greenhouse soilless culture as an alternative to conventional field cultivation. By adjusting weather, the amount and make-up of the fertilizer solution, as well as the growth media, this protected growing system can control the growing environment. As a result, when compared to traditional soil cultivation, the quality of horticultural crops developed by soilless culture is much higher .

Gaining a sufficient output that satisfies consumer demands and quality interests will be the driving force behind the future of the agricultural industry. In environmental controlled agriculture, the soilless culture system is frequently used to improve growing in open spaces and to prevent problems with the soil's water and nutritional status. The closed-loop approach now prevails in soilless culture, which previously used an open-loop method. Closed systems are proven to produce better results when it comes to maintaining crop quality and maximizing water use efficiency [35].

With the help of this artificial growing medium, plants can develop more quickly with the help of water, nutrients, and mechanical support. Hydroponics has been used commercially on occasion to grow both food and ornamental plants all around the world over the years. It is now a widely accepted methodology for plant biological research across several disciplines. Over time, the pure solutions culture has seen many changes all over the world. In a soilless culture method, sand and gravel are primarily utilized to support plants and help them store water and nutrients. Due to their distinct properties of moisture retention, aeration, leaching or action of capillary, and reusability, other substrates were subsequently developed. Comparing soilless culture media to an open field, soilless culture media is more manageable and may provide a better development environment for the plant (in terms of one or more characteristics of plant growth).

In soilless crops, plant roots can be grown either directly in nutrient solutions without any solid stage or in porous media (substrates) and irrigated often with nutrient solutions. It has recently been the standard cultural practice to fertilize plants with nutrient solutions to increase crop nutrition (fertilization or liquid fertilization). Therefore, the only characteristics of soilless crops that set them apart from crops produced in soil are their very constrained volume and uniformity of the rooting medium [40].

3.1 Necessity of Hydroponics

Because farming may be done in parched deserts and harsh environments, hydroponics can be advantageous (Turner). Due to the controlled atmosphere, growth in such situations can be accomplished via greenhouses or indoor culture. With 9.6 billion people on Earth by 2050, there will be less land available for food production [16]. To help feed the world's growing population, we will need to create new agricultural technology. Cities are growing and urbanizing the land quickly, which will help to support the research of and use of hydroponics methods since they don't require soil [16]. Around the world, hydroponic agricultural output has considerably expanded in recent years. Additionally, more effective use of fertilizers and water will be permitted, and pests and climatic conditions will be better controlled. Additionally, the production of crops using hydroponics is improving crop quality and yield, which boosts competition and economic incomes [26]. The nutrient solution is one of the most crucial predictors of crop output and quality among the parameters influencing hydroponic production systems. Soilless culture is always one that uses hydroponics, but not all soilless farming does. Less than 1/10 to 1/5 as much water as soil-based farming is used in hydroponics. Plants can be raised either in a natural nutrient solution or on sterile surfaces devoid of microorganisms [26]. The following are some reasons why soilless farming is significant: absence of soil-borne pathogens, a secure substitute for disinfecting the soil; Providing equal amounts of nourishing sandy water to plants reduces waste and brings the environment closer to ideal growing conditions; There is a chance that soilless farming will enhance production.

The ability to grow greenhouse crops with high productivity and good quality even on saline or wet soils or on soils that aren't suited for farming due to their weak structure (representing the majority of the world's arable land); A potential increase in crop yield of more than 10 times; a rise in the early yield of crops cultivated during the cold season as a result of daytime temperatures being high in the zone of roots; by environmental regulations (such as controlling or eliminating the leakage of nutrients from greenhouses into the environment and minimizing the usage of fertilizers). Therefore, building indoor soilless culture systems in greenhouses is required by law in several nations, particularly in places with controlled environments or those with restricted water supplies [30].

3.2 Benefits of Hydroponics Systems

Currently, hydroponics is a good installed type of agriculture. It has been rapid, and the results applied in many countries have proven to be practical and to have very specific benefits over traditional farming ways. The main advantage of growing plants without soil is, firstly, a much higher crop yield, and secondly, the ability to use hydroponics in locations where conventional farming or gardening are impractical. Additionally, there are additional benefits that are detailed below [34]:

- Compared to crops produced in open fields, less space is needed to produce the same amount of food, and the growth period must be shortened. Because there is no mechanical interference with the roots and all nutrients are readily available to the plant, plant growth is accelerated.
- Labor and maintenance will be reduced because the intercultural process is almost absent or very less, and fertilizing and watering are automated.
- Conservation of water is the biggest benefit. Hydroponics saves incredible quantities of water because it is using less than 1/20 of the quantity a typical farm uses to produce the same amount of production. Water logging isn't happening.
- Saving money by recycling nutrient solutions. In the case of hydroponics in the closed system, the nutrients are recycled thus preventing the loss of nutrients and avoiding soil pollution. Large volumes of water can be recycled after aeration and disposal of hypoxic conditions.
- Decreasing problems of Pest and disease became easily controlled.
- Crops that are grown in hydroponics avoid soil-borne pests.
- We can control the system and the root environment of plants like root zone temperature, humidity, darkness, etc.

- Higher productivity can be achieved because the number of plants per area is more than compared to traditional farming and the products can be achieved over a long period of crops.
- Excellent production quality, saving money, and increasing incomes.
- A Virginia hydroponic farmer has developed a head of lettuce enriched with calcium and Increasing incomes. Some plants can be enhanced through the season which can bring maximum income to the farmers.
- Besides being a commercially useful technology, hydroponics is also a standard technology used in biological research and teaching [34].

3.3 Problems of Hydroponics

Although there are several advantages of hydroponics over traditional farming, there are also some problems:

- Higher setup cost,
- Farmers need the skill and proficiency to achieve higher production in commercial applications,
- Since every plant in a hydroponic system shares the same micronutrients, diseases and pests can easily affect every plant,
- Plants have a faster reaction for changing in the environment, however, if this change is for the worse, then plants will react to it quickly; Shows signs of deficiency or trouble,
- Higher temperature and low oxygen may reduce productivity and can lead to crop losses [34].

3.4 Classification of Soilless Culture Systems

The type of soilless culture (synthetic, nutrient solution, organic medium, or a mix) and hydroponics, in which the roots are partially or submerged in a nutrient solution, are the two categories into which soilless cultures are typically classified. For several reasons—differences in the supply of nutrients throughout the system of delivery, variation in plant growth and consequent differences in nutrient uptake rate, (often scarce) irrigation water quality—The number of nutrients and water supplied must be greater than what the crop needs. All plants are properly fed by an abundance of nutrients and water, and leaching prevents excessive salt concentration and unnecessary minerals (like sodium) at the root level. Soilless systems are also classified in terms of leachate management (filtered solution) as either open-loop or closed-loop systems [40].

3.4.1 Water Culture or Hydroponic Systems

The Greek terms hydro and panic, which together indicate "working in water," are the root of the English phrase hydroponics. Instead of using dirt, hydroponics uses the nutrient solution to produce plants. The words "hydroponics," "hydroponics," "food cultivation," "soilless culture," "soilless cultivation," "pond farming," and "chemical cultivation" are all used to refer to this method of growing plants. A person who uses hydroponics is referred to as a hydroponic scientist. With or without the use of a support medium (such as sand, gravel, wormwood, Rockwool, perlite, moss, coconut, or sawdust) to provide mechanical support, hydroponics is a method for growing plants in nutrient solutions (fertilizers that contain water). There is no additional supporting medium for plant roots in hydroponic systems. Systems with a substrate have a robust support medium. Because runoff originates from treated soil and little water is lost through evaporation, it does not affect our environment. In areas affected by drought, it will be quite helpful. The very porous nature of this aqueous medium is intended to provide ideal water and air retention. Healthy breathing strengthens the roots of plants. The plants in this hydroponic system will receive top-notch, balanced nutrients [27].

Deep water culture or hydroponic systems include a bucket filled with nutrient solution, covered with a net and cloth and a thin layer of sand (1 cm) placed for supporting plants. Float hydroponics plants have been grown on foam floating in solution nutrient tanks. NFT is a Nutrient Film Technique, where a very thin nutrient solution layer is flowing in watertight channels [40].

Nutrient Film Technique (NFT)

In general, channels have a rectangular or triangular section and are lined with a variety of plastic materials, including polyethylene, polyvinyl chloride (PVC), and polypropylene. To maintain a thin fluid flow, the channel's base may be flat and not curved. The inlet flow rates vary between (1 and 3 L) per minute depending on the yield and channel size (2–$9 \, L \, m^{-2} \, h^{-1}$). For crops like lettuce, low water flow rates are preferred; for fruiting vegetables, greater rates are preferred. Additionally, a contrast can be seen between the flow rates needed for a young crop (e.g., 2–$4 \, L \, m^{-2} \, h^{-1}$) and the finished crop (e.g. 5–$9 \, L \, m^{-2} \, h^{-1}$). When flow rates fall outside of this range, they are usually often indicative of either nutrition or oxygen deficiency, or of fast flow rates that cause the water to become too deep and reduce root oxygenation; Nutrient deficits are the result of too much-delayed production, especially for plants that are downstream and exposed to water from other plants that have already absorbed minerals, particularly nitrogen and potassium. The length of the pipe affects how quickly minerals are depleted along it. As a general rule, pipes shouldn't be longer than 12–16 m. Super Nutrient Film Technology (SNFT), a modified technology, has been enhanced to prevent these issues. The nutrient solution is disseminated by a fogger installed along the channel, ensuring optimal availability of both nutrients and oxygen close to the roots [40]. A constant stream of the mineral solution is pumped into growth channels (often tubes) in Nutrient Film Technique (NFT) systems, passes through the roots of the plants, and then is discharged back into the tank. Typically, the plant's roots are hung in the nutrient solution and it is supported by a thin net or sturdy basket. NFT systems are prone to power outages, pump failures, and roots dry out quickly when the solution of nutrient flow is stopped [17]. The NFT system is designed to maintain adequate aeration of the roots by growing the plants in a thin layer of nutrient solution and can suffer from deficient O_2 concentrations due to the consumption of roots and microorganisms. NFT recycling has become the standard approach being studied for potato production [6].

A. Nutrient Solution in NFT

The nutrition film method (NFT) uses a very thin stream of water that contains all the nutrients and minerals needed for plant growth. It recycles the water from the irrigation tank while passing through the bare roots of the plants and into watertight channels. The "nutrient layer" is the term used to describe the very thin recirculating stream depths that are more than just a water coating. This ensures that, despite the presence of moisture in the air, the thick root will remain at the bottom of the canal and that there will be an adequate supply of oxygen for the plant roots. An NFT system must have a suitable channel length, slope, and flow rate to be effectively designed. The NFT system's key advantage over other types of hydroponics is that the plant roots are exposed to an adequate quantity of water, oxygen, and nutrients. These advantages lead to the maximum yield and highest-quality harvest in a little period after planting. NFT has one drawback—it can't keep water in the channel—but overall, it's one of the most successful approaches. All conventional NFT systems share the same design characteristics. Slopes between 1:30 and 1:40 are suggested along the channels. This will permit some surface irregularity, even though there may be slopes, puddles, and waterlogging. Each groove's flow rate should be 1 L/min. When we are planting, the rate may be as low as 0.5 L/min, but 2 L/min seems to be the maximum. Rates of flow that are higher than these extremes are frequently linked to dietary issues. If the channel length is greater than 10–15 m, plant growth rates won't be as rapid [34].

A continuous 24-h recirculation state is maintained for the delivery of the nutrient solution (watering and drought periods to enhance oxygen of the root system). The continuous circulation of the nutritional solution throughout the day is another alternative between these two methods (from morning to afternoon) and automated closed at night. However, if the solution of nutrient recycling is sporadic, the tank capacity should be huge enough to allow all of the solutions of nutrients included in the system when the irrigation cycle operating is shut down. Before transplanting, the channels may be covered with a 0.15–0.25 mm black on white polyethylene film, the film will be placed where the white side faces out (to reflect light and avoid intensive heat for roots and solution of nutrient) and the black side will be inside (for avoiding transmission of light and consequential on it from the evolution of algae). Plants must be modified for use in NFT systems and are placed in pots, small plastic cups, or Rockwool cubes. Once a substantial root system has developed, the cups are then inserted into the pipes [40].

B. **Advantages of Nutrient Film Technique**

The key advantages of this growing method over a static culture system are its ease of operation and the ability to regulate temperature and nutrient concentrations in a big storage tank that is feeding thousands of potential plants [34]. The advantages of NFT over other systems are the lack of substrate and the small number of solution minerals needed, which leads to significant water and fertilizer savings and lower environmental costs and substrate disposal consequences. On the other hand, because there isn't much water, the nutrient solution is subject to temperature changes along the pipe and throughout the growing seasons. Additionally, NFT has some protection against disruptions in the supply of nutrients and water. The NFT system can technically be used to plant most types of crops, but it works best with those that have a short growing season (30–50 days), like lettuce, because the plants will be ready for harvesting before the root mass fills out the canal [40].

Aeroponics

Hydroponic methods include aeroponics. The Latin roots of "aero" (for air) and "panic" are where the name "aeroponic" originates (work). Plant culture in the air is referred to as aeroponic. This kind of situation is natural [27]. The roots in this system will intermittently be fogged with fine water and nutrient solution droplets in the form of a nutrient solution fog while suspended in the air in a closed box. This approach does not call for a substrate. Plants are transplanted with suspended roots in a growth environment that regularly mists the roots with micronutrients. Excellent ventilation is one of the key advantages of the aeroponic approach. Although aeroponic farming methods are thought to be extremely effective in breeding, they have not yet been implemented on a commercial basis. Additionally, plant research in laboratories frequently employs aeroponics. NASA pays particular attention to aeroponic techniques since they are easier to apply in a zero-gravity environment where there is fog than they are with liquids [34].

Vertical Farming Systems (VFS)

The basic idea of vertical farming is to grow more food on a less amount of ground. Cultivation can be done for the same reasons that we can stack homes and offices on scarce and expensive land, like in Hong Kong or Manhattan. Supporters of the vertical farm assert that it would develop small, self-sufficient ecosystems that will serve a variety of purposes, from food production to waste management. By enabling effective and ecological food production, vertical farming will boost the economy, reduce pollution, create new jobs, rebuild ecosystems, and provide access to wholesome food. Crops will be less susceptible to the whims of the weather, pest infestation, nutrient cycle, crop rotation, pollutant runoff, pesticides, and dust in a controlled environment. As a result, indoor farming can result in better food growing in an ideal setting. Indoor farming may potentially offer higher production and more substantial returns due to its year-round operation and independence from weather conditions. Additionally, indoor farming offers a low-impact solution that can drastically save travel expenses and greenhouse gas emissions by shortening trip distances between farms and the neighborhood market. Additionally, vertical farming can boost regional economies by giving metropolitan areas much-needed "green collar" jobs [4]. To increase crop output per area unit of cultivated land, Vertical Farming Systems (VFS) have been proposed as a novel engineering technology. VFS expands crop production to vertical spaces, improving the use of land for crop production. The construction of tall food-producing structures involves the large-scale installation of vertical farming, which entails stacking growing rooms, such as greenhouses, and chambers that provide a regulated climate, on top of one another. Vertical farming systems (VFS), which operate on a smaller scale, can use the same reasoning [46].

Strawberries are grown without soil to meet the year-round increase in demand. The creation of innovative production techniques and support structures that can compete with conventional farming in terms of costs and profits will be essential for the advancement of soilless growing. The strawberry plants' growth and production are influenced by the kind of containers used and how they are arranged inside the playhouse. To produce strawberries using the greenhouse volume, a vertical production system must be set up (Verti-Gro system). However, the lower parts' subpar environmental conditions hurt plant growth and yield. Commercial aquaculture systems are used to raise high-value crops in nations including the USA, Japan, Australia, and Italy. The technology reduces the amount of global warming and promotes energy efficiency [32].

To increase crop productivity while using less land, these systems of growth extend crop productivity into the vertical dimension. Examples of VFS include the use of vertical columns, vertically suspended growth bags, conveyor-led stacked growing systems, A-frame designs, and the plant-plant method. Although crop productivity has been determined through these researches, there have been some direct comparisons with horizontal systems that use similar cropping densities, and there is little information on whether vertical column systems are a practical alternative to horizontal crop production systems [46].

Furthermore, earlier yield comparisons between VFS and conventional horizontal systems have confused the trend of crops with other parameters. For instance, compared to traditional soil cultivation, VFS has been shown to enhance output by 129–200% and profit by 3.6–5.5 US dollars/m^2. However, VFS employs a soilless growing medium, therefore invalidating the comparison. Similar to how strawberries produced in column VFS reported much higher yields than those grown in standard growth bags and multi-level VFS, but no details were supplied due to the root zone size of the growing systems [46].

3.4.2 Substrate Culture

Agriculture based on the substrate is a good replacement for the soil-based technology used in the nation. Utilizing various organic and inorganic substrates enables plants to better absorb nutrients and to grow and develop to the point where water and oxygen retention is improved. However, many substrates include a variety of chemicals that may have an indirect or direct impact on plant development [2]. Thus, selecting the optimum substrate from a variety of materials is crucial for maximizing plant productivity. The creation of substrates for soilless farming is a result of the difficulty and expense of controlling pests and diseases brought on by soil, soil salinity, a lack of fertile soil, a lack of water, a lack of space, etc. [2]. The properties of various growing media materials demonstrate direct and indirect effects on plant growth and productivity. When choosing substrates, several technical and financial variables are taken into consideration. Sand or gravel was initially utilized, but subsequently, ingredients like peat, vermiculite, and perlite became more popular. Plant protection issues with soil-borne infections and environmental rules against pesticide and nitrate contamination of groundwater are the driving forces behind replacing soil as a growing medium [2].

3.5 Coupling Between Artificial Intelligence and Soilless Culture

The Internet of Things can be used to automatically operate hydroponic systems, and ML, a subset of AI, is extremely helpful in this field of agriculture. However, less research has been done on the use of ML in hydroponic systems to automate plant development [5]. Several methods for assessing crop productivity that deviates from standard methods have recently been developed, including simulation models of process-oriented crops, statistical models for analyzing crop yields, and explanatory factors [27, 22]. Due to their easier computation and interpretation of higher power, traditional statistic-based methods or functions of specific responses that link yield and independent variables provide an alternate technique to yield prediction [36]. However, due to their limited ability to generalize to other regions and their focus on applying to local conditions, traditional empirical regressing models have certain drawbacks [10, 36]. Although ML is a "black box" with complex operations, it can manage complex interactions between independent and dependent variables [6, 23]. In recent years, ML approaches have been used in various areas of agricultural research, including crop classification, growth monitoring, and production prediction in some nations [38, 43, 47].

Because establishing a hydroponic system has a large upfront cost, it is crucial to use ML models to predict crop productivity before installing the system. The best model scenarios will assist us to define the ideal system. As a result, this may be accomplished by applying ML models to estimate the yield and weight of plants under available conditions for input scenarios consisting of combinations of input variables [31].

3.6 Interaction Between Water Management and Soilless Culture

Irrigation control ensures that the plant's identical nutrient solution supply requires items at all times, but to control plant growth (e.g., productive or vegetative) other treatments are needed. Especially in long-season crops like tomato, which is a permanently harvested crop, there must be a physiological balance between vegetative growth and generative growth. Plant treatments are necessary until the delicate flowering 15. For cucumbers, new varieties have been bred that can independently regulate the formation of fruits. There is a 'source equilibrium' for the representations of each plant between the leaves as the source and troughs, such as the plant apex, flower, fruit, and finally roots [42]. Modern greenhouse production systems can be described as either soilless or sol-based, the latter of which is usually found in arrangement processes. By creating and putting into place technologies that provide growers more control over the crucial root zone than is achievable in the soil, growers are on a trend to improve root conditions for the crop. The public views hydroponic systems as hydroponic systems without solid media and hydroponic systems as soilless production systems. The root zone, the aerial organs of plants, the irrigation system that supplies the root zone with a nutrient solution, and the drainage system that handles runoff are all parts of a hydroponics production system. In hydroponics, irrigation systems replace the various elements that are held in the root zone and then massively inflow these materials through the channel [42].

4 Conclusion

All over the world face climate change and its impacts on all countries such as decreasing of oxygen and water resources and water quality and quantity in agriculture. On the other hand, agriculture consumes the most quantities of water resources of any country. Furthermore, population growth needs more quantities of food and water. Therefore, it was necessary to work on developing new techniques of agriculture to save water and produce crops. Hydroponics is one of the new techniques to save water, increase productivity, and promote oxygen production, which is the growing of plants without soil. To get more benefits from hydroponics and save more water, hydroponics will be coupled with AI. There are many forms of coupling between hydroponics and AI such as ML, deep learning, machine vision, and the internet of things. Many types of research on ML and deep learning were conducted to predict conditions, productivity, water quality, and water use. Furthermore, the internet of things can be used in hydroponics to control greenhouse systems.

References

1. Adams, R. M., Hurd, B. H., Lenhart, S., & Leary, N. (1998). Effects of global climate change on agriculture: An interpretative review. *Climate Research, 11*, 19–30.
2. Ahmad, M. G., Hassan, B., & Mehrdad, J. (2011). Effect of some culture substrates (date-palm peat, cocopeat, and perlite) on some growing indices and nutrient elements uptake in greenhouse tomato. *African Journal of Microbiology Research, 5*(12), 1437–1442.
3. Al-Karaki, G. N., & Al-Hashimi, M. (2012). Green fodder production and water use efficiency of some forage crops under hydroponic conditions. *ISRN Agronomy, 2012*, 1–5.
4. Al-Kodmany, K. (2018). The vertical farm: A review of developments and implications for the vertical city. *Buildings, 8*(2).
5. Araújo, E. M., de Lima, M. D., Barbosa, R., & Alleoni, L. R. F. (2019). Using machine learning and multi-element analysis to evaluate the authenticity of organic and conventional vegetables. *Food Analytical Methods, 12*, 2542–2554. https://doi.org/10.1007/s12161-019-01597-2
6. Chang, D. C., Park, C. S., Kim, S. Y., & Lee, Y. B. (2012). Growth and tuberization of hydroponically grown potatoes. *Potato Research, 55*(1), 69–81. https://doi.org/10.1007/s11540-012-9208-7
7. Cireşan, D., Meier, U., Masci, J., & Schmidhuber, J. (2012). Multi-column deep neural network for traffic sign classification. *Neural Networks, 32*, 333–338. https://doi.org/10.1016/j.neunet.2012.02.023
8. Coello, C., & Lechuga, M. (2002). MOPSO: a proposal for multiple objective particle swarm optimization. In: *Proceedings of the 2002 Congress on Evolutionary Computation (CEC'02)* (Cat. No.02TH8600). Published. https://doi.org/10.1109/cec.2002.1004388
9. El-Ssawy, W., Elhegazy, H., Abd-Elrahman, H., Eid, M., & Badra, N. (2022). Identification of the best model to predict optical properties of water. *Environment, Development and Sustainability.* https://doi.org/10.1007/s10668-022-02331-5
10. Folberth, C., Baklanov, A., Balkoviè, J., Skalskı, R., Khabarov, N., & Obersteiner, M. (2019). Spatio-temporal downscaling of gridded crop model yield estimates based on machine learning. *Agricultural and Forest Meteorology, 264*, 1–15. https://doi.org/10.1016/j.agrformet.2018.09.021
11. Frank, B. (2021). Artificial intelligence-enabled environmental sustainability of products: Marketing benefits and their variation by consumer, location, and product types. *Journal of Cleaner Production, 285*, 125242.
12. Gautam, H. R., & Kumar, R. (2007). Need for rainwater harvesting in agriculture. *Kurukshetra, 55*, 12–15.
13. Gautam, H. R., & Sharma, H. L. (2012). Environmental degradation, climate change and effect on agriculture. *Kurukshetra, 60*, 3–5.
14. Gautam, H. R. (2009). Preserving the future. In *Joy of life—The mighty aqua.* Bennett, Coleman & Co. Ltd., The Times of India.
15. Hassan, K. K. (2015). *Magnetic treatment of brackish water for sustainable agriculture* (pp. 15–126). The American University in Cairo. Msc. thesis of Science in Environmental Engineering.
16. Heredia, N. A. (2014). *Design, constractionand evaluation of a vertical hydroponic tower* (pp. 5–27). BioResource and Agricultural Engineering Department. California Polytechnic State University, San Luis Obispo.
17. Hewett, E. W., & Warrington, I. J. (2014). Creation of harvesting the sun: A profile of world horticulture. *Acta Horticulturae, 1051*(14), 15–22.
18. IPCC. (1998). Principles governing IPCC work, Approved at the 14th session of the IPCC.

19. IWMI. (2008). *Helping the world adapt to water scarcity* (pp. 1–5). Internatioal Water Management Institute.
20. Indolia, S., Goswami, A. K., Mishra, S., & Asopa, P. (2018). Conceptual understanding of convolutional neural network—A deep learning approach. *Procedia Computer Science, 132*, 679–688. https://doi.org/10.1016/j.procs.2018.05.069
21. Jha, K., Doshi, A., Patel, P., & Shah, M. (2019). A comprehensive review on automation in agriculture using artificial intelligence. *Artificial Intelligence in Agriculture, 2*, 1–12. https://doi.org/10.1016/j.aiia.2019.05.004
22. Johnson, D. M. (2014). An assessment of pre- and within-season remotely sensed variables for forecasting corn and soybean yields in the United States. *Remote Sensing of Environment, 141*, 116–128. https://doi.org/10.1016/j.rse.2013.10.027
23. Kamir, E., Waldner, F., & Hochman, Z. (2020). Estimating wheat yields in Australia using climate records, satellite image time series and machine learning methods. *ISPRS Journal of Photogrammetry and Remote Sensing, 160*, 124–135. https://doi.org/10.1016/j.isprsjprs.2019.11.008
24. Kim, C. (2009). Strategies for implementing green growth in agricultural sector. In *Proceedings in Green Korea 2009—Green Growth and Cooperation*. National Research Council for Economics, Humanities and Social Science.
25. Kim, C. (2012). The impact of climate change on the agricultural sector: Implications of the agro-industry for low carbon, green growth strategy and roadmap for the East Asian Region. In *Low carbon green growth roadmap for Asia and the Pacific*. Korea Rural Economic Institute.
26. Kumari, S., Pradhan, P., Yadav, R., & Kumar, S. (2018). Hydroponic techniques: A soilless cultivation in agriculture. *Journal of Pharmacognosy and Phytochemistry, 1*, 1886–1891.
27. Lakkireddy, K., Kasturi, K., & Rao, K. (2012). Role of hydroponics and aeroponics in soilless culture in commercial food production. *Research & Reviews, 1*(1), 26–35.
28. Mall, R. K., Singh, R., Gupta, A., Singh, R. S., Srinivasan, G., et al. (2006). Impact of climate change on Indian agriculture: A review. *Climate Change, 78*, 445–478.
29. Ministry of Water Resources and Irrigation, Egypt. (MWRI). (2014). Water scarcity in Egypt: the urgent need for regional cooperation among the Nile Basin Countries.
30. Mohamed, T. M. K., Gao, J., Abuarab, M. E., Kassem, M., Wasef, E., & El-Ssawy, W. (2022). Applying different magneticwater densities as irrigation for aeroponically and hydroponically grown strawberries. *Agriculture, 12*, 819. https://doi.org/10.3390/agriculture12060819
31. Mokhtar, A., El-Ssawy, W., He, H., Al-Anasari, N., Sammen, S. S., Gyasi-Agyei, Y., & Abuarab, M. (2022). Using machine learning models to predict hydroponically grown lettuce yield. *Frontiers in Plant Science, 13*, 706042. https://doi.org/10.3389/fpls.2022.706042
32. Murthy, B. N. S., Karimi, F., Laxman, R. H., & Sunoj, V. S. J. (2016). Response of strawberry cv. Festival has grown under vertical soilless culture system. *Indian Journal of Horticulture, 73*(2), 300–303.
33. PACEST (Presidential Advisory Council on Education, Science and Technology). (2007). *Current Status and Prospects for Climate Change*. Presidential Committee on Green Growth. (2010). *Road to our future: Green growth.*
34. Pandey, R., Jain, V., & Singh, K. P. (2009). Hydroponics agriculture: Its status, scope and limitations. *Research gate*, 20–29.
35. Putra, P. A., & Yuliando, H. (2015). Soilless culture system to support water use efficiency and product quality: A review. *Agriculture and Agricultural Science Procedia, 3*, 283–288.
36. Qader, S. H., Dash, J., & Atkinson, P. M. (2018). Forecasting wheat and barley crop production in arid and semi-arid regions using remotely sensed primary productivity and crop phenology: A case study in Iraq. *Science of the Total Environment, 613–614*, 250–262. https://doi.org/10.1016/j.scitotenv.2017.09.057
37. Royal society. (2020). An overview from the Royal Society and the US National Academy of Sciences.
38. Sadeghipour, O., Aghaei, P., & Sadeghipour, O. (2013). Improving the growth of cowpea (*Vigna unguiculata* L. Walp.). *Magnetized Water, 3*, 37–43.

39. Sajith, G., Srinivas, G., Golberg, A., & Magner, J. (2022). Bio-inspired and artificial intelligence enabled hydro-economic model for diversified agricultural management. *Agricultural Water Management.* https://doi.org/10.1016/j.agwat.2022.107638
40. Savvas, D., & Gianquinto, G. (2013). Soilless culture Status report on the present situation of greenhouse crop sector in the South-Eastern European countries. Good Agricultural Practices for greenhouse vegetable crops. FAO. 217-603.
41. Schad, I., Schmitter, P., Saint-Macary, C., Neef, A., Lamers, M., Nguyen, L., Hilger, T., & Hoffmann, V. (2011). Why do people not learn from flood disasters? Evidence from Vietnam's northwestern mountains. *Natural Hazards, 62*(2), 221–241. https://doi.org/10.1007/s11069-011-9992-4
42. Schroder, F., & Lieth, J. (2002). Irrigation control in hydroponics. In *Hydroponi production of vegetables and ornamentals* (Chap. 7, pp. 263–298).
43. Shah, S. H., Angel, Y., Houborg, R., Ali, S., & McCabe, F. (2019). Spectralspatial attention networks for hyperspectral image classification. *Remote Sensing, 11*, rs11080920. https://doi.org/10.3390/rs11080920
44. Shaikh, T. A., Rasool, T., & Lone, F. R. (2022). Towards leveraging the role of machine learning and artificial intelligence in precision agriculture and smart farming. *Computers and Electronics in Agriculture.* https://doi.org/10.1016/j.compag.2022.107119
45. Srivastava, P., & Singh, R. M. (2017). Agricultural land allocation for crop planning in a canal command area using fuzzy multiobjective goal programming. *Journal of Irrigation and Drainage Engineering, 143*(6), 04017007.
46. Touliatos, D., Dodd, I. C., & Mcainsh, M. (2016). Vertical farming increases lettuce yield per unit area compared to conventional horizontal hydroponics. *Food and Energy Security, 5*(3), 184–191.
47. Wolanin, A., Camps-Valls, G., Gómez-Chova, L., Mateo-García, G., van der Tol, C., Zhang, Y., et al. (2019). Estimating crop primary productivity with Sentinel-2 and Landsat 8 using machine learning methods trained with radiative transfer simulations. *Remote Sensing of Environment, 225*, 441–457. https://doi.org/10.1016/j.rse.2019.03.002
48. Xu, J., Shrestha, A. B., Vaidya, R., Eriksson, M., Hewitt, K. (2007). The Melting Himalayas-Regional Challenges and Local Impacts of Climate Change on Mountain Ecosystems and Livelihoods. ICIMOD Technical Paper. International Centre for Integrated Mountain Development (ICIMOD), Kathmandu, Nepal.

Emerging Climate Change Technologies
in Healthcare Sector

The Influence of Climate Change on the Re-emergence of Malaria Using Artificial Intelligence

Yasmine S Moemen, Heba Alshater⬤, and Ibrahim El-Tantawy El-Sayed

1 Introduction

Natural internal processes, external forcing, and persistent anthropogenic changes in atmospheric composition and land use can all be implicated in climate change. The term "climate change" refers to a long-term, statistically significant shift in the climate's mean or variability, which can be traced to various natural causes [1].

Because of the tangled web of causality connecting malaria and global warming, there are still numerous doubts about the mechanisms at play. Climate change will increase the risk of malaria transmission in areas where the disease has previously been prevalent, where the disease has been successfully controlled, and in new areas where the disease has not previously been prevalent. Malaria-carrying mosquitoes may be increasing at higher elevations due to rising temperatures, rainfall, and humidity levels. This will lead to an upsurge in the spread of malaria in previously unrecognized areas [2]. Warmer temperatures will accelerate the parasite's life cycle in mosquitoes at lower elevations, where malaria is already a concern. This will lead to an increase in disease spread and a consequent rise in disease burden [3, 4].

Certain diseases spread by mosquitoes are associated with increased risk during the El Nio cycle. Malaria, dengue fever, and Rift Valley fever are all examples of these

Y. S. Moemen (✉)
Clinical Pathology Department, National Liver Institute, Menoufia University, Menoufia, Egypt
e-mail: yasmine_moemen@liver.menofia.edu.eg; yasmine.moemen@gmail.com

H. Alshater
Department of Forensic Medicine and Clinical Toxicology, University Hospital, Menoufia University, Shebin El-Kom, Egypt
e-mail: Heba.alshater@med.menofia.edu.eg

I. E.-T. El-Sayed
Chemistry Department, Faculty of Science, Menoufia University, Menoufia, Egypt

Y. S. Moemen · H. Alshater · I. E.-T. El-Sayed
Scientific Research Group in Egypt (SRGE), Cairo, Egypt

© The Author(s), under exclusive license to Springer Nature Switzerland AG 2023
A. E. Hassanien and A. Darwish (eds.), *The Power of Data: Driving Climate Change with Data Science and Artificial Intelligence Innovations*, Studies in Big Data 118,
https://doi.org/10.1007/978-3-031-22456-0_14

diseases. The El Nio cycle is influenced by climate change. The hatching conditions for mosquitoes can be enhanced even in dry areas by prolonged periods of heavy rainfall. Drought and increasing humidity can turn rivers into a maze of stagnant ponds, which is where mosquitoes prefer to lay their eggs [5]. By washing away mosquito breeding grounds, heavy rains can reduce the prevalence of malaria in several parts of the world. Due to El Nio-related droughts, the number of malaria cases reported in Colombia and Venezuela rose by more than one-third. Before the usage of DDT, a synthetic agricultural pesticide used in limiting the life cycle of malaria, there was a rise in the danger of malaria in Sri Lanka due to the failure of monsoons. The failure of monsoons elevated the risk of malaria by a factor of three. As a direct result of the region's abnormally heavy rainfall, malaria epidemics have recently hit several Southern African countries [5]. In 1996, western and northwestern India saw a rise in rainfall, which increased the number of malaria cases. There was less rainfall and a drop in the number of malaria cases in the same region in 1998 [6]. It is possible that the El Nio cycle's change in melanogenic capacity can contribute to epidemics of malaria in the long term.

Malaria transmission may become more difficult to track due to climate change. Variables such as population and demographic dynamics, resistance to drugs or insecticides, and human activities like deforestation, irrigation, and swamp draining all have a role in malaria transmission. In addition, climate change could have other consequences, such as increased susceptibility to malaria. For example, societal decline and economic loss could be exacerbated by health-related consequences. An individual's inability to seek early diagnosis and treatment and participation in disease control measures like insecticide spraying might contribute to a greater spread of disease. Models have shown how much more cost-effective it would be to avert climate change by reducing carbon dioxide emissions than to use other methods of reducing malaria transmission. Environmentally friendly DDT sprays, mosquito nets, and subsidies for modern combination treatments might save an estimated 78,000 lives a year if carbon emissions are reduced by one life saved each year [7].

Two of the most essential parts of the World Health Organization's plan to battle malaria since it was first established in 1992 are surveillance and preparation (WHO). Another focus has been on the eradication of risk factors for disease, such as early diagnosis and speedy and effective treatment, as well as the development of the ability to prevent and control epidemics [8]. Artemisinin-based treatments for malaria, insecticide-treated bed nets for malaria prevention, and dipsticks for malaria detection were all launched in the 1990s. These advancements, along with new initiatives like the WHO's Roll Back Malaria campaign, the Global Fund to Fight AIDS, Tuberculosis, and Malaria creation, and the inclusion of malarial indicators in the Millennium Development Goals, have led to a renewed interest in malaria prevention and treatment. All of these steps ensured that prevention, early detection, and treatment of instances, as well as capability development, were given top consideration.

The Global Malaria Action Plan includes several objectives, one of which is eradicating malaria from particular countries [9]. A handful of countries have made significant progress in eliminating malaria over the past few years. At least eight to ten countries in the elimination stage are expected to be free of locally transmitted illnesses by 2015. According to current predictions, countries in the pre-elimination stage will be eliminated by 2015. There have been no locally acquired cases in Armenia, Egypt, or Turkmenistan for more than three years and now they are in the phase of preventing the return of malaria in these countries. The Global Malaria Action Plan's objectives are in line with this. This transition from pre-elimination to nationwide elimination occurred in Azerbaijan, Georgia, Kyrgyzstan Tajikistan, and Turkey by the end of 2009, all of which are located in Europe's European Region at WHO.com [10].

This chapter is organized as follows. Section 1 presents Introduction; Sect. 2 highlights related work; Sect. 3 provides the basics and background for malaria disease; Sect. 4 describes the role of machine learning in medical applications and SubSect 4.1 display the importance of machine learning for detecting malaria disease; SubSect 4.2 presents problems and challenges for malaria detection using machine learning. Section 5 concludes this chapter.

2 Related Work

Several climates- and weather-sensitive diseases, such as malaria, are transmitted by mosquito bites. Mosquito populations can explode and spread diseases when exceptional circumstances exist, such as severe rain. Malaria is most prevalent in sub-Saharan Africa, including Burkina Faso, Mali, Niger Republic, Nigeria, Cameroon, and the Democratic Republic of Congo (DRC) [11]. Precipitation, humidity, radiation, temperature, and atmospheric pressure are all climate variables that can be affected by climate variability [12]. Much work hasn't been done on predicting malaria incidence based on climate using machine learning (ML). None of the studies have looked at these particular nations in Sub-Saharan Africa. Another research describes a decision support system based on ML that uses climate variability to divide malaria incidence into high and low target groups [13].

Malaria dynamics and distribution in central highland regions remain contested due to rising global temperatures. A 27-year study in the Chinese island province of Hainan looked at the spatiotemporal heterogeneity of malaria and its relationship to climate change [14]. Five statistical and dynamical malaria impact models were compared using bias-corrected temperature and rainfall simulations from CMIP5 climate models for three future periods (the 2030s, 2050s, and 2080s) [15].

Several studies point to the possibility that climate change could lead to an increase in the number of cases of malaria in areas with lower average temperatures and marginal transmission rates. The effect of rising temperatures in warmer locations, where conditions currently enable endemic transmission, has gotten less attention than it should have. We investigate how increases in temperature from optimal conditions (27–30 °C and 33 °C) interact with realistic diurnal temperature ranges (DTR: 0 °C, 3 °C, and 4.5 °C) to affect the ability of key vector species from Africa and Asia (*Anopheles gambiae* and *A. stephensi*) to transmit the human malaria parasite, Plasmodium falciparum [16].

In a recent study, The VECTRI model was evaluated (VECtor borne disease community model of ICTP, TRIeste) in Senegal, which attempts to understand better the link between malaria transmission and climate conditions at a national level. Vector and parasite life cycles are simulated using a grid-distributed dynamical model coupled to a basic compartmental model. An illustration of illness progression in a human host based on the SEIR model was introduced in [17].

In another work, malaria cases from 2004 to 2016 were obtained from the Chinese Center for Disease Control and Prevention as climate parameters were obtained from the China Meteorological Data Service Center. Monthly malaria cases for four *Plasmodium* species (*P. falciparum, P. malariae, P. vivax,* and other *Plasmodium*) and monthly climate data were collected for 31 provinces. The researchers only looked at the big picture. They didn't use any private information in that research [18], where the re-emergence of malaria cases was predicted using the long short-term memory sequence-to-sequence (LSTMSeq2Seq).

3   Basics and Background for Malaria Disease

More than one billion people across Africa, Asia, and Latin America are infected with malaria, the world's most deadly tropical mosquito-borne parasitic illness. Every year, the disease takes the lives of approximately 1 million people [19]. Every United Nations member state has agreed to a set of goals known as the Millennium Development Goals (MDGs). It will be much easier to achieve these objectives if malaria damage is minimized [19]. The mosquito's lifespan, the mosquito's development of malaria parasites in the mosquito, and the subsequent transmission of malaria are all affected by changes in climate, such as those in temperature, rainfall patterns, and humidity. Due to the fact that mosquitoes serve as the principal vector for the spread of the illness [20]. The earth's average temperature has risen throughout the last century, and since the middle of the 1950s, the rate of warming has increased noticeably. Climate change has been linked to a rise in malaria, among other things [21]. This rise in mosquito-borne disease transmission is predicted to positively impact the spread of the disease and its geographic dispersion [22]. Among the possible effects of climate change is a rise in malaria. Malaria has been controlled or eradicated in formerly endemic areas in some studies. Still, other researchers have found no link between climatic change and the disease's spread [23–25] or re-emergence in previously endemic areas in other studies [26, 27]. Other studies report that there is no association between malaria and climate change [28]. Malaria was once widespread throughout Europe, notably in Scandinavian countries. As a result of better social and economic conditions, better irrigation and drainage, new farming practices, behavioral shifts, and better health care, the illness was eradicated from Europe in 1975. This was accomplished despite the Earth's rising average temperature [29, 30].

4 Machine Learning for Medical Applications

Medical professionals can profit from using ML algorithms in decision-making by identifying effective therapies and optimum measures. A support vector machine (SVM) has been used in prior studies [31] to forecast the occurrence of malaria based on climate data, such as in an early warning system (SVM). Malaria incidence can be predicted using climatic data, and from the free weather and geography Application Programming Interface (API), you can get information about temperature, relative humidity, wind speed, sun radiation and rain [32]. In a study done in China, ML algorithms were used that use a mix of many algorithms to finish tasks and predict performance better than any single algorithm could. With the stacking method of generalisation error reduction they combined predictions from different primary learning algorithms, such as temperature, humidity and vapor pressure, as well as daily rainfall and moisture levels to predict whether malaria would be positive or negative based on meteorological variables such as wind speed and length of sunshine [33]. As well as a daily report on clinical data, Thakur et al. [34] employed the vegetative index, rainfall, and relative humidity to create an artificial neural network model for anticipating the occurrence of malaria in India. Following research, scientists could make an educated guess as to how many malaria cases there will be in the following years. In some parts of the world, ML algorithms can predict when an outbreak of malaria will occur, according to their research. In the face of climate unpredictability, no single model can be relied upon as the gold standard for malaria forecasting because each method has its unique modeling norm that is dependent on the behavior of the research area. Figure 1 summarises comparative analyses of the available methods.

In Fig. 1, Multiple ML methods are used to examine the influence of climate change on malaria transmission. For, **SARIMA** [35, 36], the model successfully predicted the seasonal trend in malaria using periodic or seasonal time-series data, even though it is only appropriate for stationary or seasonal processes. **ARIMA** [37], using moving averages, accurately predicted the number of malaria cases in Afghanistan. It can also detect and quantify underlying patterns' influence in time-series data. Predicted time series data can be easily forecasted using the ARIMA model. Non-seasonal data cannot be accurately predicted using this method; **Binomial Model** [38], In the negative binomial model, climate factors and the rate of malaria transmission were accurately connected. There was no consideration of the relationship between climate variables and transmission rates in any other part of Nigeria or Africa. Data trends can be difficult to discern because this study relies on a statistical approach; **VECTOR** [39], predicts irregularities in the transmission of malaria in Uganda using statistical methods. The model uses a statistical model to anticipate the exact numbers of malaria transmission in Uganda, which may not be perfect for future predictions; **SLIM** [40], captures uncertainty in data and allows malaria to be disrupted by the use of this method. The statistical model used in this study solely looked at the relationship between the malaria vector and a climatic

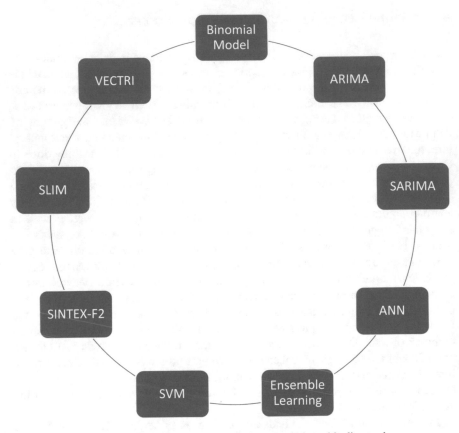

Fig. 1 Different ML models used to evaluate malaria transmission with climate change

variable; **SINTEX-F2** [41], An interpretable and flexible model was used to charac-terise the non-linear, delayed association between weather and malaria cases. This study only looked at precipitation and temperature as two climate variables. As a result of these uncertainties, the SINTEX-F2 model's effects on malaria cases could be misclassified; **SVM** [31], An investigation into the links between several meteo-rological variables was carried out using the PLS-PM approach. Malaria outbreaks may be successfully predicted with the use of ML techniques. The SVM model provided the most accurate predictions. The SVM model performs better in high-dimensional spaces when there is a clear margin of separation between the target classes; **Ensemble Learning** [32], Their findings imply that ensemble models outper-form standard time-series models in terms of performance; **ANN** [33], the most accurate stacking framework for malaria prediction may be challenging to design due to the lack of consideration of all feasible models. A ML technique was utilised to uncover complex nonlinear relationships between meteorological variables such as temperature, rainfall, relative humidity, and vegetative index for the forecast of

Khammam district in India. First and foremost, projecting the precise levels of malaria could lead to erroneous predictions in light of illness prevalence.

VECTRI model driven by reanalysis data was used to simulate Senegal's entomological inoculation rate (EIR) (ERA-5). Climate Prediction Center (CPC) data from Africa Rainfall Climatology 2.0 (ARC2) satellites and Climate Hazards InfraRed Precipitation with Station data are all used to supplement the daily ERA5-Land reanalysis rainfall. These other meteorological products include the CPC Global Unified Gauge-Based Analysis of Daily Precipitation (CPC) (CHIRPS). PNL/Programme national de lutte contre le paludisme au Sénégal) information on the frequency of malaria in the country [17].

Another ML model used data from 2004 to 2016 to predict the re-emergence of malaria cases using a long-short term memory sequence-to-sequence deep neural network model. The XGBoost, gated recurrent unit, LSTM, and LSTMSeq2Seq models were trained and tested using monthly malaria cases and supporting meteorological data from 31 Chinese provinces. It was then determined how accurate the models' predictions were, using root mean squared error (RMSE) and mean absolute error (MAE) [18].

This long-term cohort study studied the relationship between temperature and malaria transmission in Hainan, China, utilizing a decades-long dataset of malaria incidence records from China's Hainan. The climate data was compiled using data from the WorldClim dataset and local meteorological stations in Hainan. Researchers used a temperature-dependent R0 model and a negative binomial generalized linear model [14] to better understand the relationship between climatic conditions and malaria prevalence in tropical locations.

Both global and regional analyses were conducted on climate suitability, the additional population at risk, and additional person months at risk. Projections for malaria use five global climate models, each with four emission scenarios (Representative Concentration Pathways, RCPs) and one single population forecast. The modeling uncertainty of climate change-related estimates of malaria-risk populations was also addressed. According to our research, global climate appropriateness and the population at risk are expanding despite the many uncertainties [15].

4.1 The Importance of Machine Learning for Detecting Malaria Disease

In a recent study [17], The climatic data analysis results were compared. Rainfall in Senegal's latitudinal gradient may be seen in the country's seasonal malaria transmission disparity, which is strongly linked to the country's rainfall. With a one-month delay between rainfall and the peak of malaria in Senegal, September and October are peak months for mosquitoes. Based on observations and computer calculations, malaria cases are expected to drop over time. According to these studies, malaria outbreaks are more prevalent in Senegal's southern region. Thanks to the data offered in such studies, it will be easier to tailor national malaria prevention, response, and care activities to the specific requirements of Senegalese communities [17].

In the latest research [18], the comparison between proposed models, the suggested LSTMSeq2Seq model reduced the mean RMSE of the predictions for *P. falciparum, P. vivax, P. malariae*, and other plasmodia by 19.05–33.93%, 18.4–33.59%, 17.6–26.67%, and 13.28–21.34%. The average prediction accuracy for the LSTMSeq2Seq model was 87.3% [18].

According to previous work [13], Using real-world data, a new intelligent system can classify shifts in malaria prevalence according to climate variance. Climate changes appear to impact malaria incidence in the six countries evaluated in Sub-Saharan Africa. Thus, the climate attribute that has the greatest effect on malaria varies from country to country. Climate variables such as rainfall and solar radiation were found to have a significant impact on malaria transmission in all six study locations. Feature engineering, k-means, and hyperparameter optimization were all employed to improve the MIC model's precision. Using the findings of previous studies [13, 18], It's possible to properly plan for future malaria epidemics if decision-makers have more information. Each country's government will be better able to identify the environmental factors that lead to high malaria transmission rates as a result of this method, which would reduce the prevalence of malaria in that country by addressing those causes. This approach. Budgetary considerations can be eased by the spread of insecticide-treated nets and malaria medicines as part of efforts to eliminate the disease [13]. Secondly, to improve the models' capacity to predict, a larger dataset is needed, especially for cases of malaria that have been verified and such a dataset should have a resolution equivalent to or finer than the climate observations used in the training. Data on verified malaria incidence, possibly in the form of time series data that might seasonally stratify crucial seasons for real-time prediction, is an important aspect of the future effort.

According to new research, an annual peak in Hainan has been found to occur earlier in the central highlands and later at lower elevations this year. Model results show that temperatures of about 15 °C have been linked to long-term changes in incidence peak time. There is a 95% confidence interval for the decrease in malaria incidence if the temperature rises 1 °C from the northern plains to the central highland regions during the rainy season by 56–92%. Forty-six percent (95% CI 37–55%) to 119% of the population shifts from low- to high-altitude areas during the dry season (95% CI 98–142%) [14]. So, such a study [14] recommended that rising temperatures can have opposite effects on the dynamics of malaria in low- and high-lying areas. Future modeling, disease burden estimations, and malaria control in central highland regions under climate change should consider this.

The model's projections predict a net increase in the number of yearly person-months at risk when comparing Representative Concentration Pathways (RCP) 2.6 to RCP8.5 throughout the period 2050–2080 [15]. Over the epidemic edges of the malaria distribution, the result measures were particularly sensitive to whether the malaria impact model was used [15]. The evaluation of the malaria output models for the two observed climate baseline datasets and the GCM modeled baselines with other published malaria endemicity maps based on [42] to try and validate the various malaria models. For preintervention malaria endemicity estimations, Lysenko used historical records, papers, and maps to compile malariometric indexes for the four most common *Plasmodium* species (*malariae, vivax* and *falcipárum*) [43]. It has been determined that *P. falciparum* endemicity (MAP) values can be used for comparison. With the help of survey data, environmental and socioeconomic factors are combined to provide a "best guess" of the global prevalence of the disease, which is then used to map it. From the original papers, these datasets have been digitized. This can be compared to the outputs of the malaria model, which describe the epidemic and stable transmission zones based on LTS.

In another study, with optimal conditions, the temperature range of both mosquito species saw a reduction in their total vectorial ability due to the impacts of increased temperature and DTR on the prevalence and intensity of parasites as well as the death rate of mosquitoes. Increases in temperature of 3 °C from 27 °C reduced the vectorial capacity by 51–89%, depending on the species and DTR. Increases in DTR alone had the potential to cut transmission in half. At a temperature of 33 °C, the transmission potential of the insect *Anopheles stephensi* was further reduced, whereas it was inhibited in *Anopheles gambiae* [16]. These findings suggest that even slight temperature variations could play a significant part in the dynamics of malaria transmission, yet very few empirical or modelling investigations take this into account. In addition, they predict that current and future warming may, rather than raise risk, lower transmission potential in existing high transmission situations [16].

4.2 Problems and Challenges for Malaria Detection Using Machine Learning

Till Now, there isn't a vaccine available for malaria, so malaria transmission continues even lags one to two months behind seasonal rainfall. Senegal's rainy season is July–August–September, although the malaria season is September–October–November. Temperature also affects malaria.

To simulate climate-influenced malaria transmission, the VECTRI model has to be tested. An entire month between the EIR peak (September) and rainfall can be simulated by using the yearly cycle (August). EIR and malaria cases corresponded in both years of high and low malaria transmission. The VECTOR can predict some of the seasonal fluctuations in malaria. Like the observation data, the EIR indicates Senegal's south-north direction to be unevenly distributed. In the north, centre, and south, this was evident.

In fall season (September, October, and November), the simulation predicted a high rate of malaria transmission. Despite irregularities, the VECTRI model accurately depicted the geographic and temporal distribution of malaria in Senegal. To better understand how climate changes affect the spread and transmission of malaria, researchers can use this model. The VECTRI [17] has the potential to improve vector-borne disease early warning systems. In the next step, the VECTRI model was utilised to see how well CORDEX-driven simulations could reproduce historical malaria characteristics over Senegal and quantify projected changes under RCP.45 and RCP8.5. The WHO came up with the idea for it. According to the authors, researchers in West Africa and Senegal hope to examine regional climate model downscaling and the multi-model ensemble in conjunction with VECTRI to better understand the timing and degree of regional malaria change. A bias-corrected CMIP5 dataset would be used to simulate malaria parameters for several RCP scenarios and other periods in the VECTRI model Decision-makers in countries affected by climate change, such as Senegal, will certainly benefit from these findings when developing public health efforts. Stakeholders can use these findings to develop vector-control methods [17]. Malaria re-emergence was accurately predicted using the LSTMSeq2Seq model, which considered climatic conditions. As a result, the LSTMSeq2Seq model [18] can be used to accurately forecast the onset of malaria.

Due to a lack of daily malaria incidence records, a previous study [13] only considered annual data to predict seasonal climatic parameters. Second, to improve the models' predictive power, a larger dataset is needed, particularly for cases of malaria that have been verified, with observations having a similar or finer resolution to climate observations for training. There is a need for a good dataset for verified malaria incidence that can be used to seasonally stratify critical malaria seasons to improve real-time prediction in the future.

5 Conclusion

Malaria epidemic planning will benefit from this research in the future. A key benefit of this approach is that it will help the governments of each selected country better understand the climatic parameters that contribute to high malaria transmission and better regulate environmental factors that could hurt climate conditions. An increase in funding for programs like sensitization, the provision of insecticide-treated nets, or malaria treatments can be achieved.

References

1. Climate change 2007: Impacts, adaptation and vulnerability. Genebra, Suíça (2001).
2. Jetten, T. H., Martens, W. J. M., & Takken, W. (1996). Model simulations to estimate malaria risk under climate change. *Journal of Medical Entomology, 33*(3), 361–371.
3. Rogers, D. J. (1996). Changes in disease vector distributions. In *Climate change and Southern Africa: An exploration of some potential impacts and implications in the SADC Region* (pp. 49–55). Climate Research Unit, University of East Anglia, Norwich.

4. Sutherst, R. W. (1998). Implications of global change and climate variability for vector-borne diseases: Generic approaches to impact assessments. *International Journal for Parasitology, 28*(6), 935–945.
5. WHO. (2002). Fact sheet 192: El Niño and its health impact.
6. Krishnamurti, T. N., Chakraborti, A., Mehta, V. M., & Mehta, A. V. (2007). Experimental prediction of climate-related malaria incidence. *Atelier sur la mousson ses conséquences.*
7. Lomborg, B. (2009). *On climate advice to policy makers.* The Copenhagen Consensus.
8. Roll Back Malaria. (2005). *World malaria report 2005.* World Health Organisation, UNICEF.
9. Roll Back Malaria. (2008). Roll Back Malaria partnership. *A Global Malaria Action Programme.*
10. Otten, M., Cibulskis, R. E., Williams, R., WHO Global Malaria Programme, & Aregawi, M. (2009). *World malaria report 2009.* World Health Organization.
11. Centers for Disease Control and Prevention. (2019). Malaria's impact worldwide. Global Health, Division of Parasitic Diseases and Malaria. https://www.cdc.gov/malaria/malaria_w orldwide/impact.html. 26 Jan 2021.
12. Tompkins, A. M., & Thomson, M. C. (2018). Uncertainty in malaria simulations in the highlands of Kenya: Relative contributions of model parameter setting, driving climate and initial condition errors. *PLoS ONE, 13*(9), e0200638.
13. Nkiruka, O., Prasad, R., & Clement, O. (2021). Prediction of malaria incidence using climate variability and machine learning. *Informatics in Medicine Unlocked, 22,* 100508.
14. Wang, Z., et al. (2022). The relationship between rising temperatures and malaria incidence in Hainan, China, from 1984 to 2010: A longitudinal cohort study. *The Lancet Planetary Health, 6*(4), e350–e358.
15. Caminade, C., et al. (2014). Impact of climate change on global malaria distribution. *Proceedings of the National Academy of Sciences, 111*(9), 3286–3291.
16. Murdock, C. C., Sternberg, E. D., & Thomas, M. B. (2016). Malaria transmission potential could be reduced with current and future climate change. *Science and Reports, 6*(1), 1–7.
17. Fall, P., Diouf, I., Deme, A., & Sene, D. (2022). Assessment of climate-driven variations in malaria transmission in Senegal using the VECTRI model. *Atmosphere (Basel), 13*(3), 418.
18. Kamana, E., Zhao, J., & Bai, D. (2022). Predicting the impact of climate change on the re-emergence of malaria cases in China using LSTMSeq2Seq deep learning model: A modelling and prediction analysis study. *British Medical Journal Open, 12*(3), e053922.
19. Vasant, N., & Amir, A. (2003). "Roll back malaria?" The scarcity of international aid for malaria control. *Malaria Journal, 2*(1), 1–8.
20. Hoshen, M. B., & Morse, A. P. (2004). A weather-driven model of malaria transmission. *Malaria Journal, 3*(1), 1–14.
21. Stark, J., Mataya, C., & Lubovich, K. (2009). Climate change, adaptation, and conflict: A preliminary review of the issues. In *United States Agency for International Development—53 Climate Change Conference. Conflict-Sensitive Climate Change Adaptation in Africa.*
22. Gubler, D. J., Reiter, P., Ebi, K. L., Yap, W., Nasci, R., & Patz, J. A. (2001). Climate variability and change in the United States: Potential impacts on vector-and rodent-borne diseases. *Environmental Health Perspectives, 109*(suppl2), 223–233.
23. Zhou, G., Minakawa, N., Githeko, A. K., & Yan, G. (2005). Climate variability and malaria epidemics in the highlands of East Africa. *Trends in Parasitology, 21*(2), 54–56.
24. Zhou, G., Minakawa, N., Githeko, A. K., & Yan, G. (2004). Association between climate variability and malaria epidemics in the East African highlands. *Proceedings of the National Academy of Sciences, 101*(8), 2375–2380.
25. Pascual, M., Ahumada, J. A., Chaves, L. F., Rodo, X., & Bouma, M. (2006). Malaria resurgence in the East African highlands: Temperature trends revisited. *Proceedings of the National Academy of Sciences, 103*(15), 5829–5834.
26. Baldari, M., et al. (1998). Malaria in Maremma, Italy. *Lancet, 351*(9111), 1246–1247.
27. Krüger, A., Rech, A., Su, X., & Tannich, E. (2001). Two cases of autochthonous *Plasmodium falciparum* malaria in Germany with evidence for local transmission by indigenous *Anopheles plumbeus. Tropical Medicine and International Health, 6*(12), 983–985.

28. Hay, S. I., et al. (2002). Climate change and the resurgence of malaria in the East African highlands. *Nature, 415*(6874), 905–909.
29. Kuhn, K. G., Campbell-Lendrum, D. H., & Davies, C. R. (2002). A continental risk map for malaria mosquito (Diptera: Culicidae) vectors in Europe. *Journal of Medical Entomology, 39*(4), 621–630.
30. Füssel, H.-M., Klein, R., & Ebi, K. (2006). Adaptation assessment for public health. In *Climate change and adaptation strategies for human health* (pp. 41–62).
31. Kim, Y., et al. (2019). Malaria predictions based on seasonal climate forecasts in South Africa: A time series distributed lag nonlinear model. *Science and Reports, 9*(1), 1–10.
32. Modu, B., Polovina, N., Lan, Y., Konur, S., Asyhari, A. T., & Peng, Y. (2017). Towards a predictive analytics-based intelligent malaria outbreak warning system. *Applied Sciences, 7*(8), 836.
33. Wang, M., et al. (2019). A novel model for malaria prediction based on ensemble algorithms. *PLoS ONE, 14*(12), e0226910.
34. Thakur, S., & Dharavath, R. (2019). Artificial neural network based prediction of malaria abundances using big data: A knowledge capturing approach. *Clinical Epidemiology and Global Health, 7*(1), 121–126.
35. Paolella, M. S. (2018). *Linear models and time-series analysis: Regression, ANOVA, ARMA and GARCH*. Wiley.
36. Adeola, A. M., et al. (2019). Predicting malaria cases using remotely sensed environmental variables in Nkomazi, South Africa. *Geospatial Health, 14*(1).
37. Mopuri, R., Kakarla, S. G., Mutheneni, S. R., Kadiri, M. R., & Kumaraswamy, S. (2020). Climate based malaria forecasting system for Andhra Pradesh, India. *Journal of Parasitic Diseases, 44*(3), 497–510.
38. Anwar, M. Y., Lewnard, J. A., Parikh, S., & Pitzer, V. E. (2016). Time series analysis of malaria in Afghanistan: Using ARIMA models to predict future trends in incidence. *Malaria Journal, 15*(1), 1–10.
39. Baghbanzadeh, M., et al. (2020). Malaria epidemics in India: Role of climatic condition and control measures. *Science of the Total Environment, 712*, 136368.
40. Tompkins, A. M., Colón-González, F. J., Di Giuseppe, F., & Namanya, D. B. (2019). Dynamical malaria forecasts are skillful at regional and local scales in Uganda up to 4 months ahead. *Geohealth, 3*(3), 58–66.
41. Le, P. V. V., Kumar, P., Ruiz, M. O., Mbogo, C., & Muturi, E. J. (2019). Predicting the direct and indirect impacts of climate change on malaria in coastal Kenya. *PLoS ONE, 14*(2), e0211258.
42. Gething, P. W., Smith, D. L., Patil, A. P., Tatem, A. J., Snow, R. W., & Hay, S. I. (2010). Climate change and the global malaria recession. *Nature, 465*(7296), 342–345.
43. Lysenko, A. J., & Semashko, I. N. (1968). Geography of malaria. A medico-geographic profile of an ancient disease. *Itogi Nauki: Medicinskaja Geografija, 25*, 146.

Index

© The Editor(s) (if applicable) and The Author(s), under exclusive license
to Springer Nature Switzerland AG 2023
A. E. Hassanien and A. Darwish (eds.), *The Power of Data: Driving Climate Change with
Data Science and Artificial Intelligence Innovations*, Studies in Big Data 118,
https://doi.org/10.1007/978-3-031-22456-0

Printed in the United States
by Baker & Taylor Publisher Services